NORTH CAROLINA
STATE BOARD OF COMMUNITY COLLEGES
LIBRARIES
SAMPSON COMMUNITY COLLEGE

FIRE SERVICE ENTRANCE EXAM PREPARATION

STUDENT MANUAL

BY

ARTHUR R. COUVILLON

and

PAUL H. STEIN

INFORMATION GUIDES, HERMOSA BEACH, CALIFORNIA

FIRE SERVICE ENTRANCE EXAM PREPARATION

STUDENT MANUAL

1991

BY

ARTHUR R. COUVILLON

and

PAUL H. STEIN

FIRST EDITION
ALL RIGHTS RESERVED.
THIS BOOK, OR PARTS THEREOF,
MAY NOT BE REPRODUCED IN ANY FORM
WITHOUT PERMISSION OF THE PUBLISHER
Printed in the United States of America
COPYRIGHT, 1991 BY INFORMATION GUIDES

Library of Congress Cataloging Data:

Couvillon, Arthur R. and Stein, Paul H.
"Fire Service Entrance Exam Preparation": Student Manual
1. Fire Service: Handbooks, manuals, text books, etc.
2. Fire Engineer Handbooks, manuals, text books, etc.
3. Fire Captain: Handbooks, manuals, text books, etc.
4. Firefighter: Handbooks, manuals, text books, etc.
5. Information: Handbooks, manuals, text books, etc.
6. Promotional: Handbooks, manuals, text books, etc.
I. Title
LCCN 91-073745
ISBN 0-938329-71-5

INTRODUCTION

The objective of this book is to help you increase your chances of obtaining a career in the Fire Service. This book we will discuss all aspects of the Fire Service entrance examination process. In the chapters to come we will zero in on the various components of the exam including the written, physical abilities and oral interview.

This book will include diagnostic exams for you to take. The purpose of these exams is to identify any deficiencies you might have regarding various exam topics and to identify means for you to improve in these areas. Upon successful completion of this book and if you follow all the instructions and advice given, you will improve your chances tremendously to obtain a career in the Fire Service.

The intent of this book is to provide a list of information pertaining to the duties, knowledge, responsibilities, qualifications, training, and education that is necessary in order to successfully become an **ENTRANCE FIREFIGHTER** in the Fire Service.

The goal of this book is to show **ENTRANCE FIREFIGHTER** candidates a program that they may use to successfully begin a career in the Fire Service.

ACKNOWLEDGEMENTS

Appreciation is expressed to those individuals, Rookie Firefighters, Fire Engineers, Fire Captains, Fire Training Officers, Fire Technology Instructors, Fire Chiefs, and Personnel Directors that have contributed to the compilation of this system for "FIRE SERVICE ENTRANCE EXAM PREPARATION".

ABOUT THE AUTHORS

Paul H. Stein, is a Battalion Chief with the City of Santa Monica. Chief Stein has been in the Fire Service for 22 years and has been involved in teaching at the college level for 15 years. Additionally, Chief Stein is the Fire Technology coordinator at Santa Monica College, and an instructor for the California Fire Service Training and Education System. In 1983 Chief Stein was honored by the California Fire Chief's Association as the California Fire Service Instructor Of the Year.

Arthur R. Couvillon is a veteran Firefighter, with 20 years of Firefighting experience. Art shares the knowledge that he has gained during this period with a series of published Fire Service Study guides:

THE COMPLETE FIREFIGHTER CANDIDATE

FIREFIGHTER WRITTEN EXAM STUDY GUIDE

FIREFIGHTER ORAL EXAM STUDY GUIDE

ADVANCEMENT IN THE FIRE SERVICE

FIRE ENGINEER WRITTEN EXAM STUDY GUIDE

FIRE ENGINEER ORAL EXAM STUDY GUIDE

FIRE CAPTAIN WRITTEN EXAM STUDY GUIDE

FIRE CAPTAIN ORAL EXAM STUDY GUIDE

All books are available from:
INFORMATION GUIDES, P. O. Box 531, Hermosa Beach, CA 90254

TABLE OF CONTENTS

CHAPTER #1 :
INTRODUCTION TO THE FIRE SERVICE
PAGES: 1 - 32

CHAPTER #2 :
CAREER OPPORTUNITIES AND
THE APPLICATION PROCESS
PAGES: 33 - 65

CHAPTER #3 :
THE WRITTEN EXAMINATION
PAGES: 67 - 131

CHAPTER #4 :
THE ORAL INTERVIEW
PAGES: 152 - 174

CHAPTER #5 :
MATH CONCEPTS AND ARITHMETIC
PAGES: 175 - 205

CHAPTER #6 :
AREAS, VOLUMES, CAPACITIES,
AND BASIC HYDRAULICS
PAGES: 207 - 221

CHAPTER #7 :
READING COMPREHENSION/
IMPROVING READING SKILLS
PAGES: 223 - 233

CHAPTER #8 :
READING COMPREHENSION/
CONCENTRATION AND MEMORY
PAGES: 235 - 275

CHAPTER #9 :
BASIC SCIENCE/CHEMISTRY/PHYSICS
PAGES: 277 - 299

CHAPTER #10 :
MECHANICAL COMPREHENSION
PAGES: 301 - 316

CHAPTER #11 :
MECHANICAL APTITUDE
PAGES: 317 - 355

CHAPTER #12
FIRE SERVICE INFORMATION
PAGES: 357 - 377

INDEX:
PAGES: 379 - 386

CHAPTER 1

INTRODUCTION TO THE FIRE SERVICE

FIRE SERVICE CAREERS/OPPORTUNITIES

INTRODUCTION:

The objective of this chapter is to increase your chances of obtaining a career in the Fire Service. This chapter will examine all aspects of the Fire Service entrance exam process. The chapters to follow will delve into the various components of the exam including the written, physical abilities and the oral interview.

This book includes various diagnostic exams for you to take. The purpose of these exams is to identify any deficiencies you might have regarding various exam topics and to identify means for you to improve in these areas. Upon the completion of this book and the following of the instruction and advice given, you will improve your chances tremendously to obtain a career in the Fire Service.

FIRE SERVICE CAREER CLASSIFICATIONS:

PAID FIREFIGHTER: A current member of a paid Fire Department. May be private, industrial, or any of the above mentioned services.

VOLUNTEER FIREFIGHTER: A normally enrolled Firefighter that dedicates time and application to a community Fire Service without any type of pay or compensation, except for death and injury benefits such as Workman's Compensation.

CALL FIREFIGHTER: A part time Firefighter, may be paid by the hour or with an annual retainer, or both.

EXPLORER/CADET FIREFIGHTER: A Firefighter trainee, Pre Fire Academy type training.

PRIVATE BUSINESS FIREFIGHTER: A full time paid Fire-Protection specialist with responsibility for Fire Prevention/Protection needs of a private facility.

FIRE DEPARTMENT: A fire protection organization, usually of municipal or county government, or private industry/business that provides fire extinguishment, fire prevention, fire protection, and emergency rescue services to a given jurisdiction or area.

PAID FIRE DEPARTMENT: A Fire Department that is usually located in the larger Cities. They are usually comprised of a large staff that depends upon Paid Firefighters for the day-to-day operations, with adequate off-duty personnel available for a major incident or disaster.

PARTIALLY-PAID/PARTIALLY VOLUNTEER FIRE DEPARTMENT: Are usually located in areas that can not provide a large enough paid-full time reserve firefighting force. By the use of Volunteer Firefighters as a reserve force, a jurisdiction may reduce their budgets. Well trained Volunteer Firefighters are capable of providing excellent support to the full-time Paid Firefighting force.

VOLUNTEER FIRE DEPARTMENT: usually in an area that cannot pay for necessary basic level of fire protection. Volunteer Firefighting organizations may not furnish the same level of service that a Paid Fire Department, although Volunteer Firefighters are normally highly motivated with great concern and interest in their organization.

VARIOUS FIRE SERVICE POSITIONS

FIRE SUPPRESSION:

FIREFIGHTER: this is the entrance level position of an active member of a Fire Department, paid or volunteer.

ACTING POSITION, ACTING ENGINEER, ACTING OFFICER: an individual of a lower rank detailed to the duties and responsibilities of a higher rank.

FIRE ENGINEER, APPARATUS DRIVER, PUMP OPERATOR: the person assigned to a fire apparatus (pumper, aerial ladder, boat, etc.) and is responsible for driving and operating such apparatus.

FIRE LIEUTENANT: the first officer of a Fire Company, in full command when the Fire Captain is not present, and usually in command of one company.

FIRE CAPTAIN, COMPANY COMMANDER: the first level of supervision, the officer in charge of a company or Fire Station. Usually the highest ranking officer of an individual company, the basic organizational unit of the Fire Department, which consist of a team of personnel, under the direct supervision of the Fire Captain.

CHIEF FIRE OFFICER: the highest ranking individual at a fire scene, usually the highest ranking suppression officer.

BATTALION CHIEF, DISTRICT CHIEF, B/C, D/C: a chief officer just below the rank of Deputy or Assistant Chief and immediately above the rank of Fire Captain. This is an intermediate supervisory level. The Battalion Chief usually commands a number of fire companies.(five to eight companies are normal)

DIVISION CHIEF: a Deputy Chief Fire Officer in command of a division, entire shifts, or platoons of personnel, which consist of several Fire Department battalions.

DEPUTY CHIEF, ASSISTANT CHIEF: a high ranking Chief Officer of a Fire Department, usually second or third in the chain of command below the Fire Chief Of the fire Department.

FIRE CHIEF, CHIEF ENGINEER: The highest ranking and commanding firefighting officer of a given Fire Department. Chief Engineer is the former designation for the Chief of the department when the Chief's responsibility was the direction of engineers and care of engines. In some cities Chief Engineer is still the title of the Fire Chief.

ADMINISTRATIVE OFFICERS: these are the executive-level managers that endure the ultimate responsibility for managing the Fire Department.
Administrative Officers consist of the Fire Chief, Deputy Chiefs, and Assistant Chiefs.

FIRE SUPPORT SERVICES:

FIRE PREVENTION BUREAU: the unit of the Fire Service that is involved in Fire Prevention and Fire Investigation, including inspection and code compliance endeavors, induction of new codes and ordinances along with plan checking.

TRAINING DIVISION: includes a Training Officer that is responsible for the training and education of entrance Firefighters; evaluation of the Fire Departments competence; testing and implementing of new techniques, methods, and procedures; conducting research, and carries on continuous training of all personnel.

WATER ENGINEERING: includes a Water Supply Officer that is responsible for working with water authorities on design and placement of hydrants; also works with property developers to secure an adequate and reliable supply of water for any given district; maintains contact with suppression and notifies suppression upgrades and/or any changes of water supply and if necessary of supplementary water supplies.

DISPATCH OFFICER: for the development of a smooth, efficient system that will allow for quick receipt of alarms and the competent and effective deployment of personnel and equipment to incidents; includes Fire Dispatchers that receive alarms, confirm locations, an responds appropriate personnel and equipment to the incident.

FIRE/ARSON INVESTIGATION: this involves the investigation of every fire, including a systematic examination of the fire scene along with questioning of witnesses to determine the cause of the fire. Also additional investigation of all suspicious fires so as to ascertain the innocence or guilt of wrong doing of those involved.

PUBLIC INFORMATION AND EDUCATION: principal purpose is to generate understanding of the significance of fire and inform the public as to what they can do to reduce fires frequency and severity; educates the public as to how to practice Fire Prevention; informs the public about Fire Department operation and activities.

ADMINISTRATIVE SUPPORT: includes additional support operations needed for the effective administration of a contemporary Fire Department. Examples:
1. Apparatus and automotive maintenance/repairs.
2. Budget and equipment inventory.
3. Personnel administration.
4. Clerical support.
5. Data processing.
6. Payroll.
7. Research and development.
8. Internal affairs.

FIRE SERVICE WORKING CONDITIONS

The majority of Fire Departments continue to work the traditional 24 hour shifts/56 hour week, in suppression, which allows the Firefighters to have numerous days off duty. This is mostly a positive benefit to the Firefighters, with the exception of days off following a shift with responses to emergencies during the sleeping hours. Be aware that there are other schedules in the Fire Service. Make sure that you know the working schedule of the Fire Department that you are applying!

Firefighters may use these days off to advance their education, pursue hobbies, participate in recreational activities, relax, or even "C shift": work additional jobs, start a business. ETC!

The best way to describe the working conditions of the Fire Service is with the term **"ESPRIT DE CORP"**:
1. Team work (including peer pressure)
2. Comradeship
3. Second family
4. Friendships

Because of the 24 hour working schedule Firefighters eat and sleep together along with their normal job requirements and responsibilities.

Firefighters gain a great deal of job satisfaction with their abundance of diverse opportunities and task that occur during a 24 hour shift, such as:
1. Responding to and extinguishing both major and minor fires.
2. Responding to and rectifying both major and minor "Haz-Mat" incidents.
3. Responding to and aiding victims/patients in need of emergency medical attention, whether it be a minor traffic accident to a major accident requiring extrication along with major traumatic injuries or a patient in full arrest.

Firefighters may respond to anything from the most minor incident to the most serious type of incident that you might imagine!

During some of the "Working-Hours" (usually 0800-1700 hours) of a 24 hour shift, Firefighters may be involved in:
1. Training/drills.
2. Company fire prevention inspections and pre fire plans.
3. Public relations.
4. Physical fitness programs.
5. Hydrant maintenance/testing.
6. Apparatus and equipment maintenance/testing.
7. Station/grounds maintenance.
8. Shopping and meals.
9. Related duties as required.
10. **EMERGENCY RESPONSES!**

During some of the after work hours (usually 1700-0800 hours) of a 24 hour shift Firefighters may utilize this time for:

1. Recreation/games.
2. Personal hygiene.(shower and comfort facilities are provided)
3. Meals.(Firefighters are responsible for providing and preparing their own meals, this may be done by individuals providing their own food, or an arrangement where all in station Firefighters buy and prepare the food together.
4. Self studying
5. Pleasure reading or television.
6. Sleep.(usually dormitory type facilities)
7. **EMERGENCY RESPONSES!, ETC!**

Some of the benefits included in a career as a Firefighter are:

1. Job security. (especially in the civil service status that exist in many municipalities and cities.)
2. Promotional advancement opportunities.
3. Wages are competitive with the business/industry world.
4. Vacation/holiday/sick pay.
5. Retirement plans.
6. Family: health-medical/dental/vision plans.
7. Life/accident insurance.
8. Uniform/shoes/equipment allowances.
9. Pay for unused sick leave.
10. Industrial injury leave.
11. Deferred compensation plans.
12. Longevity pay and Educational bonuses.
13. Premium pay for special assignments.
14. Periodic medical examinations.
15. Personal leave.
16. Disability benefits.
17. On duty physical exercise programs.
18. Military leave.

ALL OF THE ABOVE BENEFITS ARE SUBJECT TO INDIVIDUAL FIRE DEPARTMENT COLLECTIVE BARGAINING AGREEMENTS!

SUMMARY

It is important to remember the goals identified at the beginning of the book and this chapter. This book will be directed towards achieving these goals. The essential component to reach the goals is your cooperation and commitment.

Your cooperation and commitment can be measured through understanding and by trying to do your best to improve. Remember, this book is solely designed to help you achieve a career as a Firefighter.

The work world of a Firefighter is multifaceted. A person can join the Fire Service and serve as a explorer/cadet, volunteer, partially paid or fully paid. The Fire Department rank structure usually goes from Firefighter to Fire Chief with many levels in-between. Promotions are gained through competitive examinations. The Firefighter job consists of many duties from cleaning the Fire Station to fighting a multi-alarm fire. One of the most important things a prospective Firefighter can do to help his/her chances of getting on a Fire Department is to learn as much as possible about the Firefighters job.

ASSIGNMENT:

Read : "Information Sheets" 1-4:

1. Requirements and Qualifications, pages 9 - 12.
2. The Position of Firefighter, pages 13 - 17.
3. Fire Service as a Career, pages 18 - 19.
4. Probation, pages 20 - 31.

Prepare for:

1. Quizzes.
2. Class material.
3. Assigned reading.

INFORMATION SHEET #1

TOPIC : REQUIREMENTS AND QUALIFICATIONS

INTRODUCTION :
During the period that you are preparing for a career in the Fire Service you should be aware of the entrance requirements.

The following list of REQUIREMENTS and QUALIFICATIONS for entrance Firefighters has been compiled from various Fire Departments across all of the U.S.A.

Be aware that each Fire Department will have their own particular REQUIREMENTS and QUALIFICATIONS and that this list is a compilation of several Fire Departments of which some will demand several on the list, while other Fire Departments will demand only a few.

INFORMATION:
This list is only a guide to make YOU aware of many of the possible REQUIREMENTS and QUALIFICATIONS that may be needed in order to enter a Fire Service career.

EDUCATION:
1. Equivalent to graduation from High School. (GED will be accepted in most cases)
2. There is no minimum educational requirement.
3. 45 semester hours with a C average or better from an accredited College or University.

AGE:
1. 18 years of age upon graduation of basic training.
2. At least 21 years of age by final filling date.
3. Age limit of 31/34/35 years of age at interview date.
4. No age limit.
5. Must be at least 18 years old but not yet 35 years old by the last day for filling. Anyone who reaches the age of 35 while on the eligible list will have his/her name removed from the list. By state law, no one can be offered this position who has reached the age of 35 years.

MEDICAL:
1. Vision: 20/40 uncorrected with both eyes.(the weakest 20.70; the stronger 20/40). Each eye corrected to 20/30.
2. Vision no worse than 20/40, corrected to 20/20 in both eyes.
3. Standard visual acuity: without correction, less than 20/40 in one eye, and 20/100 in the other eye; and with correction, less than 20/20 in one eye, and 20/40 in the other eye.(as you can see they vary from one Department to another Department).
4. Hearing: not more than 10% loss in either ear.
5. Must have a standard hearing average threshold without correction of no worse than 40 decibel loss on the average at 500, 1000, and 3000 Hertz frequencies in the best ear.

6. Must pass body fat percentage analysis.
7. Sound mental and physical health.
8. Height and weight in proportion.
9. Physical ability to perform job functions.
10. Candidates for this position must have age, height and weight standards as established by the Civil Service Board.
11. Physical strength and agility and freedom from serious physical defects as indicated by a physical examination.
12. Medical examination meeting or exceeding NFPA 1001 standards.
13. You must be in good health and have the ability to perform the strenuous duties of a Firefighter without medical restrictions, or hazard to yourself, co-workers, or the public.
14. Physical condition must be adequate for the performance of the duties of a Firefighter as determined by the City Physician.
15. All applicants must pass a physical examination given by the County Health Department physicians and must be certified as being capable of meeting the physical, medical, and mental requirements established by the F/C.
16. Color blindness is not acceptable.
17. Those candidates finishing within certification range will be required to submit to a medical screening process at the City's expense. This medical screening process will be conducted by a licensed physician and will be used to evaluate a candidate's capability to safely perform the medically related physical demands of the job.
18. You must be physically and emotionally sound and free from any condition which might adversely affect the performance of fire fighting duty.
19. You must have normal hearing and color vision.
20. You will be tested in several aspects of physical ability such as agility, strength, coordination, balance, and stamina.

LICENSES:
1. Valid class 1 driver's license upon employment.
2. U.S. Citizenship.
3. Must possess a class "C" drivers license.
4. Will be required to obtain a class "B" drivers license during probation.
5. Possession of a valid license to legally operate any class of vehicle or equipment necessary to perform the duties of the position is required at the time of appointment.
6. State Firefighter I certificate.
7. State Firefighter II certificate.
8. County certified Paramedic.
9. Emergency Medical Technician I certificate.
10. Valid third class drivers license.
11. You must have an honorable discharge if you were in the military service.

ABILITIES, EXPERIENCE AND TRAINING:
1. Any combination of course work, experience, and training which provides the required knowledge, skills, and abilities.
2. Must be able to obtain and maintain a minimum of state EMT-1 certification.
3. Paramedic.
4. Must be able to operate or learn to operate fire apparatus within first year of employment.
5. Must be able to pass the entrance written exam.
6. Must be able to pass the physical agility testing process.
7. Completion of a State Board of Fire Service approved Basic Firefighter I Training Academy. Certificate will be required.
8. Ability to read and write.

9. Some knowledge of the street system and physical layout of the City, or ability to acquire this knowledge.
10. Ability to understand and follow oral and written instructions.
11. Ability to work long hours under discipline.
12. Ability to learn a wide variety of firefighting duties and methods, including the operation of firefighting apparatus, equipment and tools.
13. Bi-lingual capabilities, English along with such languages as Spanish, Vietnamese, or another language depending upon the ethnic make-up of Fire Departments location.
14. Graduation from a Basic Fire Academy, or experience, education equivalent to completion of a certified Fire Academy.
15. Applicants must pass all portions of the selection process: Written Exam, Physical Agility, Oral Interview, Complete Medical Exam, Psychological Exam, and Background Investigation including a drug/alcohol screening test.
16. Applicant must pass a background investigation to demonstrate the ability to satisfactorily perform firefighting duties associated with protecting the publics safety under difficult conditions/situations.
17. Must be able to pass a background investigation which will verify the accuracy and completeness of statements contained on the application and obtain information relevant to job performance.
18. All applicants shall be proven by investigation to be of good reputation, character, and morals.
19. Applicant must achieve a score of 70% or better on the Oral Interview.
20. Ability to read, write, and speak English.
21. Applicant must achieve qualifying score in the Firefighter qualifying Written Exam, or be able to meet a waiver condition.
22. Applicant must achieve qualifying scores in each of the Physical Abilities Test events.
23. Applicants shall be able to demonstrate the ability to fight fires which is a very strenuous occupation requiring above average strength and endurance. It requires a high degree of both upper and lower body strength, cardiovascular endurance and stamina, flexibility and balance. Must be able to perform under difficult environmental conditions including heights, extreme heat, darkness, confined spaces, toxic fumes, and long hours without rest.
24. There is no experience requirement. All new Firefighters will be given complete training in firefighting methods and techniques immediately after they are hired.
25. Candidates for the position of Recruit Firefighter are not required to have specialized prior training. While not mandatory experience in the following areas will benefit the candidate in both the selection process and in Drill School: Mechanical Aptitude, Building Construction, Chemistry, Emergency Medicine, and Public Relations.
26. Ability to work cooperatively with others.
27. Applicants will have successfully completed one year as a County Probationary Firefighter or three years experience as a sworn employee of the County Fire Department; plus a current certificate as a Firefighter II.
28. Aptitude for mechanical work.

SPECIAL REQUIREMENTS:
1. Must be a non-smoker; persons who have smoked during the last 12 months need not apply.
2. Women are encouraged to apply.
3. Copies of required certificates must be attached to applications to be accepted.
4. Lateral-entry Firefighters from other Fire Departments will periodically be accepted. (must meet all requirements for the position)

5. Many Fire Department Chiefs have a personal criteria for Firefighter recruits such as:

ATTITUDE, If the attitude is upbeat, can do, and pleasant, as opposed to negative, most other problems can be solved.

HONESTY, RESPONSIBILITY, and DEPENDABILITY are key personality characteristics for a successful career as a Firefighter. These traits may be reflected by participation in school or social organizations and an absence of conflict with the law along with a solid work history. Due to the combination of physical and mental stress inherent to this profession, self motivation and spontaneous problem-solving are also important factors. Outstanding INTEGRITY, COURAGE, INITIATIVE, and a NEAT PERSONAL APPEARANCE.

CHARACTER: you must have a satisfactory employment record and a good credit rating. You must not have been addicted to Barbiturates or Narcotics or excessive use of Alcohol. You must not have been convicted of any serious crime or have any criminal charges pending against you.

6. Must be a residence of the City/Area in which Fire Department is located.(at least 1 year)
7. Must be a bona fide City resident at time of interview and maintain residency until the completion of a probationary period.
8. Must have worked with the City as an Auxiliary Firefighter for a minimum of six months and successfully have passed a proficiency exam.
9. You must pass a reading exam.
10. Credit to promotion is available to Firefighters who have completed College degrees.
11. Candidates must be willing to work 24 hour shifts involving weekends and Holidays.
12. Minimum height of 5 feet 6 inches.
13. Minimum weight of 140 pounds, in proportion to height.

SPECIAL WAIVERS: Candidates who meet at least one of the following requirements will not be required to take the Written Exam:
1. Completion of 60 semester or 90 quarter units, with at least a "C" (2.0) grade point average. Units must have been earned at a College or University accredited by one of the members of the council on Post Secondary Accreditation.
2. Successful completion of an employment period as a City Fire Department Trainee.
3. Both the Written Exam and the Waiting Period for test scheduling will be waived if you have successfully completed an employment period as a City Fire Department Trainee. (as above)
4. Upon verification of completion from The Fire Department Trainee Program, your application will be accepted and you will be scheduled for testing on the next available test date.
5. Documentation will be required before any waiver will be granted.

YOU MUST MAKE YOURSELF AWARE OF THE QUALIFICATIONS AND REQUIREMENTS OF THE FIRE DEPARTMENT THAT YOU ARE INTERESTED IN HAVING A CAREER WITH!

INFORMATION SHEET #2

TOPIC : THE POSITION OF FIREFIGHTER

INTRODUCTION :
THE FOLLOWING LISTS ARE SAMPLES OF HOW VARIOUS FIRE DEPARTMENTS DESCRIBE THE POSITION OF FIREFIGHTER, ANY ONE FIREFIGHTING POSITION MAY NOT INCLUDE ALL OF THE DUTIES LISTED, NOR DO THE LISTED EXAMPLES INCLUDE ALL TASK WHICH MAY BE FOUND IN POSITIONS OF THIS CLASS!

EXAMPLE #1

A Firefighter performs firefighting and Fire Prevention activities under hazardous and stressful conditions. Typical duties include:
1. Responds with fire company to fire or emergency alarms.
2. Works at fire scene wearing air-pack, mask and protective clothing.
3. Carries, places and climbs ladders for rescue or ventilation operations.
4. Searches burning buildings for persons who may be inside.
5. Administers first aid and CPR.
6. Walks or crawls through smoke-filled areas.
7. Removes victims from burning buildings.
8. Connects hoses to water sources.
9. Carries hose bundles into buildings and up stairways.
10. Performs clean-up operations after the fire has been extinguished.
11. Cleans and maintains fire equipment and apparatus.
12. Performs inspections and fire prevention activities.
13. Performs other related duties as required.

EXAMPLE #2

Firefighter is the entrance level in the uniformed ranks of the Fire Department. An employee in this class is normally trained during the first year to perform the basic duties of the class. However, skills and abilities substantially increase as the employee acquires more training and experience in fire suppression, rescue, emergency medical service, and Fire Prevention operations. Although a Firefighter normally works in accordance with standard operating procedures and under the orders of a Fire Captain, the employee frequently makes independent decisions affecting life and property. Employees in this class must be capable of intense physical exertion for extended periods under hazardous conditions. They also must have skill in establishing and maintaining good employee and public relations, in acquiring substantial of building construction, and in interpreting legal codes. Firefighter duties include:
1. Engage directly in firefighting, Fire Prevention, emergency medical service and related activities for the saving of life and property.
2. Operates and maintains firefighting apparatus, tools and equipment.
3. May operate an assigned ambulance.
4. Firefighters may operate the smaller, less complex, fire apparatus or an ambulance, or may tiller an aerial ladder truck.

5. Normally a Fire Engineer drives, operates, and maintains heavy pumping apparatus and aerial ladder trucks, however, a Firefighter must qualify to operate such equipment and may operate these vehicle types under emergency and non-emergency conditions while the regular assigned operator is temporarily absent.
6. When assigned to a Fire Company or to the Fire Prevention Bureau, a Firefighter makes fire inspections that require less technical background, knowledge, and specialized training than is required of a Fire Inspector.
7. Responds to alarms with the company.
8. Connects and lays hose lines.
9. Enters burning buildings with hose lines.
10. Assist in rescuing persons from burning buildings and from other places where life is in danger.
11. Renders emergency medical service including Paramedic assistance.
12. Uses hand fire extinguishers and a variety of hand tools and gasoline or hydraulic-powered equipment.
13. Raises, lowers, and climbs ladders.
14. Ventilates buildings to release smoke, heat, and gases.
15. Follows established procedures for handling and making notification about hazardous materials or conditions.
16. Uses self contained breathing apparatus.
17. Assist in overhauling operations to ensure that the fire is completely extinguished.
18. Replaces broken or ruptured Fire Sprinkler System heads to prevent unnecessary water damage.
19. Protects property by spreading salvage covers and by using sawdust, brooms, mops, shovels, and similar equipment.
20. May tiller an aerial ladder truck or operate and maintain a tank wagon, squad, or smaller automotive apparatus.
21. May drive an ambulance and maintain its equipment.
22. Cleans firefighting apparatus and equipment.
23. Assist in maintaining fire station cleanliness.
24. Participates in station security procedures.
25. Attends, conducts, and participates in training activities.
26. Assist with office duties.
27. Qualifies to operate heavy pumping and aerial ladder apparatus.
28. Uses a hose line or brush tools to control grass and brush fires.
29. Inspects property for proper brush clearance.
30. Enforces fire prevention laws and regulations in the mountain fire district.
31. Stands lookout while a fire boat maneuvers.
32. Assist in extinguishing fires in vessels and on the water front.
33. Assist in extinguishing fires in aircraft.
34. Acts as company fire prevention inspector for residential and small commercial structures within an assigned district.
35. Issues notices of Fire Code Violations or Fire-Life Safety Violations to property owners.
36. Discusses violations with the public to obtain voluntary abatement of hazards.
37. Prepares drawings for building inventory program.
38. Participates in company firefighting inspections.
39. Maintains a Chief Officers vehicle and drives under non-emergency and emergency conditions.
40. Assist in varied administrative duties, overtime hiring, timekeeping, personnel procedures, and supply control.
41. At emergencies, gathers information for and forwards directions from a Chief Officer.
42. Transmits and receives radio reports.
43. Provides situation status and resource status information.
44. Compiles incident summaries from subordinate units reports.

45. May photograph unusual fire conditions or suspicious circumstances.
46. Responds to calls for emergency medical treatment or paramedic service.
47. Rescues trapped people.
48. Applies cardio-pulmonary resuscitation techniques.
49. May assess patient condition, establish a communication link to a paramedic base facility, and carry out a physician's directions in administering medications, medical treatment and other paramedic procedures.
50. May transport patients to a hospital.
51. Receives telephone calls for fires and request for emergency medical service and dispatches fire apparatus and rescue ambulances to emergency incidents.
52. Assist in controlling the distribution of apparatus and receives reports and notifies stations of apparatus out-of-service, water shut-off at fire hydrants, and closed streets.
53. Notifies the news media of unusual events.
54. Records locations and movements of Chief Officers, Arson Investigators, Photographers, and mechanics.
55. Operates complex radio switching and dispatch telephone facilities, teletype and microfiche equipment and computer terminals.
56. Directs and assists managers and trains employees in industrial and commercial facilities, high-rise structures, hospitals, and government facilities to develop and implement an emergency organizational structure and plans.
57. Presents talks and gives demonstrations to schools, civic, service, social, and other groups to increase public awareness toward fire safety.
58. Writes news releases and provides liaison with news media concerning Department fire and life safety activities and policies.
59. Assist an Incident Commander by providing information to the news media.
60. Evaluates fire hydrant locations.
61. Works with other agencies involved with hydrant installation and water supply.
62. Maintains master fire hydrant maps and records.
63. Test fire hydrants.
64. May qualify to maintain breathing apparatus.
65. May qualify for assignment as Arson Investigator, Helicopter Pilot, or "Scuba Diver" Firefighter.
66. May occasionally be assigned to other duties for training purposes or to meet technological changes or emergencies.
67. May be assigned related duties as required.

EXAMPLE #3

Firefighters perform fire fighting duties; raise and climb ladders; uses chemical extinguishers, bars, hooks, lines and other equipment; ventilates burning buildings, removes persons from danger, throws salvage covers and removes debris.

Drives various types of firefighting automotive equipment and operates pumps, aerial ladders, and auxiliary equipment.

Checks engines daily and inspects pumping equipment, ignition, batteries, brakes and other equipment to see that the apparatus is in good working condition and notifies supervisors of any defect.

Assist in repair and maintenance of Fire Department apparatus and equipment.

Participates in first aid and rescue work.

Checks and cleans equipment after returning from a fire.

Participates in fire drills, attends fire classes in firefighting, first aid, and other relevant subjects necessary for the efficient operation of the Fire Department.

Inspects all types of buildings and draws plot plans of buildings.

Participates in Fire Prevention programs to educate the public in action that can be taken to prevent fires.

Assist in keeping fire station in a clean and orderly condition.

Participates in other related work as required.

EXAMPLE #4

The Firefighter classification is the general duty firefighting work in the protection of life and property by combating, extinguishing and preventing fires.

Primary responsibility is for specialized firefighting duties under emergency conditions which may involve personal hazard.

Specific orders and directions are given by superior officers, but the work requires a thorough individual understanding of firefighting methods gained by training and experience.

A large portion of time in the fire service is spent in training and routine work in the maintenance of firefighting equipment, apparatus and quarters, inspection of premises for hazards, and the performance of varied details in and around the fire station.

Firefighters may be assigned to operate and maintain a major piece of firefighting equipment or apparatus employing special skills learned on the job.

Firefighters:

Respond to fire alarms with a fire company, enters burning buildings with hose lines, operates nozzles and plays a stream of water on fire as directed.

Remove persons from danger and gives first aid to injured persons.

Raise, lower and climb ladders and ventilates burning buildings to carry off smoke and gases.

Operate hand fire extinguishers, chemical hose, fog nozzles and similar equipment in extinguishing fires.

Perform salvage operations such as throwing salvage covers, sweeping water and removing debris.

Performs routine housekeeping duties at a fire station, such as mowing grass, removing snow, making beds, cleaning walls, windows, floors and prepares meals.

Stand watch.

Do repair and maintenance work employing special skills.

Drive pump and ladder trucks, operates pumps, aerial ladders and auxiliary apparatus, services and maintains automotive and pumping equipment.

Inspect all types of buildings for fire hazards and location of exits and the location of fire protection devices.

Prepare records of inspections; notes and reports other types of inspections; carries on continuous fire prevention activities.

Inspect fire hydrants for freezing; acts as safety officer at school crossings; acts as fire guard at large public assemblies.

Take command of a fire company in the absence of a superior officer.

Attend instruction sessions in such subjects as firefighting methods, equipment operation, first aid, Fire prevention methods, and street and hydrant locations.

Sells refuge bags as required.

Performs related work as required.

FOR A COMPLETE LIST OF FIREFIGHTERS SPECIFICATIONS - QUALIFICATIONS REQUIREMENTS REFER TO N.F.P.A. PAMPHLET 1001: "FIREFIGHTER PROFESSIONAL QUALIFICATIONS"

INFORMATION SHEET #3

TOPIC : FIRE SERVICE AS A CAREER

INTRODUCTION:
As a prospective Firefighter in your preparation for a career as a Firefighter you should be aware of the different types of Fire Department jobs available (see start of this section).

Be aware of the opportunities to improve yourself and the Department along with performing the duties and responsibilities of a Firefighter:
1. Maintenance of:
    Station.
    Fire hydrants.
    Alarm boxes.
    Apparatus and equipment.
2. Respond to emergency incidents.
3. Perform emergency procedures.
4. Perform rescue techniques.
5. Public relations:
    Tours.
    Demonstrations.
6. Take part in:
    Training.
    Fire prevention.
    Pre-fire planing.
7. Be available and ready to act as Fire Engineer.
8. Continue studying.
9. Related duties as needed.
10. Firefighter will be able to identify:
    Organization of the Fire Department.
    Size and scope of the Fire Departments operations.
    Standard Operating Procedures of the Fire Department.
    Rules and regulations.
11. Display confidence in performing your job including:
    Fire Station maintenance.
    Apparatus and Equipment maintenance.
    Firefighting.
    Company fire prevention inspections and Pre-fire planning.
    Related duties as required.
12. Firefighters should be aware of NFPA 1001 Firefighter professional qualifications.
13. Set aside time for self study every shift that you are on duty.
14. Make it obvious to your peers and supervisors that you are pleased to be a Firefighter.
15. Make yourself available for special details, assignments and projects including:
    Hydrant maintenance.
    Pump and Hose testing.
    Training and instruction.
    Tours and Demonstrations.
    Equipment fabrication.
16. Do the best job that you are capable of doing.

17. Come up with new ideas for improving your Fire Department.
18. Bid for as many varied assignments that your Fire Department makes available:
    Engine pumper companies.
    Truck companies.
    Salvage companies.
    Rescue companies.
    Paramedics.
    Dispatch.
    Fire prevention.
    High incident areas.
    Harbor districts.
    High rise areas.
    ETC.
19. Gain knowledge in:
    Fire apparatus, tools and equipment.
    Fire pumps and streams.
    Fire prevention.
    Fire behavior.
    Fire hydraulics and water supply.
    Hazardous materials.
    Fire extinguishing systems.
    Automatic and mutual aid. (response assistance)
20. Gain experience in:
    Fire Department responses.
    Driving apparatus.
    Pumping fire pumpers.
    Report writing.
21. Performing at emergency incidents:
    Hazardous material.
    Fires.
    Rescues.
    Entrapment.
    ETC.

Learn the duties of your Fire Departments next two ranks up from your present job classification, such as:
  Fire Engineer, Apparatus Driver, Pump Operator.
  Fire Captain, Company Officer, Lieutenant.

Participate in your Fire Departments certification program that allows you to "act" in those positions.

Continue to establish, add to and complete the list that was started while preparing for a career in the Fire Service.

Be aware of the testing procedures for the next promotional exam.

Study specifics for the Fire Departments next promotional exam.

INFORMATION SHEET #4

TOPIC : PROBATION

INTRODUCTION :
The final period in the selection process for ENTRANCE LEVEL FIREFIGHTER is the "PROBATIONARY PERIOD", and is just as important as any other phase of the selection process. If you are not performing adequately you may be terminated without recourse.

The PROBATIONARY PERIOD is "on the job training" with pay. During this period you will be scrutinized and evaluated as to your performance of the task and duties required for the position.

You will be evaluated not only during the performance of the duties but also during drills and training for these duties, which usually will include a period of time in a Fire Service Training Academy.

During the PROBATION you will be given daily, weekly, and monthly test to rate your progress.

PROBATIONARY REQUIREMENTS
(FOR TYPICAL AVERAGE SIZE FIRE DEPARTMENT)

FIRE DEPARTMENT MONTHLY TRAINING SCHEDULE FOR PROBATIONARY FIREFIGHTERS:

Each probationary Firefighter shall be trained on a monthly basis as outlined in the following material. The Firefighter's training shall consist of manipulative evolutions as well as various classroom drills. The Fire Captains shall make every attempt to cover the material outlined. The Fire Captains may add additional areas of training during a given month, but they shall insure that the required training is adequately covered. The Training Officer shall closely monitor the training.

The Fire Captains shall administer a comprehensive written quiz at the end of each month (except the fifth, ninth, tenth, eleventh, and twelfth) based on the material covered during the month. The Fire Captains shall forward a copy of each completed quiz to the Training Officer for review and filing.

FIRE DEPARTMENT POLICY AND PROCEDURE

SUBJECT: TRAINING, TESTING AND EVALUATING PROBATIONARY FIREFIGHTERS

I. Purpose
   A. To thoroughly train, test and evaluate probationary Firefighters.
   B. To standardize training, testing and evaluation of probationary Firefighters.
II. Policy
   A. Probationary Firefighters shall attain minimum passing grades on all major test.
   B. Probationary Firefighters shall maintain a satisfactory rating on all performance evaluations.
   C. Department officers shall administer the training, testing and evaluation process as required.
III. Responsibility
   A. The Training Officer shall develop and monitor major test and shall establish minimum training standards and guidelines.
   B. The Fire Captains shall administer the training, testing and evaluation of the probationary Firefighters.
   C. The Training Officer shall oversee and review the routine training, testing and evaluation of probationary Firefighters.
   D. The probationary Firefighter shall fully participate in training and study and prepare for the various exams.
IV. Procedures
   A. The test administered to each probationary Firefighter shall be as follows:
      1. Monthly quizzes (written) administered by Fire Captains.
      2. Written major tests administered by the Training Officer at the end of the fifth (5th) month of employment and at the end of the ninth (9th) month.
      3. A final practical exam at the end of the tenth (10th) month.
   B. The major written test will include the following subjects:
      1. Fire behavior.
      2. Fire extinguishers.
      3. Ropes and knots.
      4. Ladders.
      5. Fire streams and hose practices.
      6. Forcible entry.
      7. Rescue.
      8. Ventilation.
      9. Breathing apparatus.
      10. Salvage and overhaul.
      11. Sprinkler systems.
      12. Firefighter safety.
      13. Fire prevention practices.
      14. Emergency Medical Technician (emergency care and transportation of the sick and injured).
      15. Department rules and regulations.
      16. Department policies and procedures.
   C. The final practical exam will cover the following areas:
      1. Breathing apparatus.
      2. Hose and ladder practices and evaluations.
      3. Truck company operations.
      4. Emergency Medical Technician procedures.

D. The Fire Captains shall administer the routine training exercises based on the schedule outlined by the Training officer:
 1. Monthly quizzes (written) will be administered by the Fire Captains at the end of each month.
 2. Quizzes will be based on the month's training.
 3. Completed quizzes shall be forwarded to the Training Officer.
E. Each probationary Firefighter shall be evaluated by their Fire Captain at the end of each month using the standard evaluation form.
 1. An oral counseling session shall be made a part of the evaluation process.
F. The Training Officer will recommend the final action on probationary Firefighters to the Fire Chief.
 1. The final recommendation will be based on the final exams, practical exam and the evaluations of the Fire Captains.

MONTHLY TRAINING SCHEDULE

MONTH THREE:

I. Department Familiarization.
   A. Department organization.
      1. Organization chart.
      2. Stations.
      3. Companies.
      4. Response patterns.
      5. Platoon operations.
   B. Department rules and polices.
      1. Familiarization with key areas.
      2. Duties and responsibilities.

II. Personal safety.
   A. Safety clothing.
   B. Taking hydrant.
   C. No hose clamp.
   D. Hand signals.
   E. Flaking hose.
   F. Pre-connect lines.
   G. Evaluated drills.

III. Equipment locations on various apparatus.

IV. Paramedic/ambulance familiarization.
   A. Assisting paramedics in the field.
   B. Ambulance procedures.

V. Basic Truck Company operations.
   A. Setting up and operating aerial ladder.

VI. IFSTA Essentials: Chapter 10 and Chapter 4.

MONTH FOUR

I. E.M.T. Procedures.
   A. CPR review.
   B. Resuscitator.
   C. Primary, secondary survey.
   D. M.A.S.T.
   E. Hare traction splint.
   F. Burn packs.
   G. Gurney operations.
   H. Miscellaneous procedures.

II. Power equipment: starting and operating.
   A. Hurst tool.
   B. Saws: Circular and chain.
   C. Smoke ejector.

III. Maps.
   A. City boundaries.
   B. Major streets.
   C. Address system.

IV. Ladders.
   A. Roof.
   B. 24 foot extension.
   C. 35 foot extension.

V. Aerial ladder.
   A. Ground operations.
   B. Basket operations.
   C. Elevated stream.

VI. Review Rules and Regulations/Policies and Procedures.

VII. IFSTA Essentials, Chapter 5 and Chapter 7.

MONTH FIVE
(Mid-term exam at the end of the Month)

I. Fire Prevention.
   A. Codes, review.
      1. U.B.C.
      2. U.F.C.
      3. National Fire Codes.
   B. Reference manuals, how to use.
   C. Inspection procedures.

II. Review hose evolutions.

III. Review ground ladders.

IV. Review aerial operations.

V. Review for exam.

VI. IFSTA Essentials, Chapter 17.

MONTH SIX

I. Fire Extinguishers.
   A. Types.
   B. Practical application.

II. Ropes and knots.
   A. Selected knots.

III. Forcible entry.
   A. Tools.
   B. Types of doors and windows.
   C. Techniques.

IV. Review Rules and Regulations/Policies and Procedures.

V. IFSTA Essentials, Chapters 2, 3, and 6.

MONTH SEVEN

I. Ventilation.
   A. When needed.
   B. Tools.
   C. Types of roofs.
   D. Techniques.
   E. Safety.

II. Salvage.
   A. Covers.
   B. Throws.

III. Overhaul.
   A. Protecting, preserving evidence.
   B. Extinguishing hidden fires.
   C. Safety.

IV. Review hose evolutions.

V. Review E.M.T., Ambulance Procedures.

VI. IFSTA Essentials, Chapter 11 and Chapter 12.

MONTH EIGHT

I. Strategy and Tactics.
   A. R.E.C.E.O. (V., S.)
   B. Fire behavior.

II. Water Supply.
   A. Distribution and storage in city system.

III. Sprinklers.
   A. Fire Department support.

IV. Fire Prevention Review.
   A. Codes.
   B. Inspection procedures.

V. IFSTA Essentials, Chapters 1, 8, and 16.

MONTH NINE
(Final written exam at the end of the Month)

I. Review for final exam.

II. Review any other manipulative training that Captain may deem necessary.

MONTH TEN
(Practical exam at the end of the Month)

I. Review for practical exam.

FIRE DEPARTMENT
MANIPULATIVE FINAL EXAM:
(Probationary Firefighter)

Each probationary Firefighter will take a manipulative final exam at the end of the tenth month of employment. Each candidate will be tested in ten areas, selected at random from the following twenty areas. The Firefighter will be required to successfully pass at least eight of the areas for a passing grade.

1. Lay a supply line (wet), including taking a hydrant and making the supply line to pumper connection.
2. Flake and extend 250 feet of 2 1/2 inch working line (ground level), wet).
3. Put on breathing apparatus.
4. Place, raise and lower 24 foot extension ladder (one man).
5. Advance a hose line via 24 foot extension ladder to the second floor of the training tower (wet).
6. Extend a 2 inch pre-connect line to the third floor of the training tower via the inside stairway (wet).
7. Place, raise and extend the 35 foot extension ladder with two other personnel allowed to assist.
8. One man fold and carry (2 1/2 inch hose).
9. Explain the operation and start the Hurst power tool.
10. Explain the operation and start the chain saw and circular saw.
11. Explain the operation and start the light plant on the Ladder Truck, including setting-up lights and extension cords.
12. Perform ground operations for setting-up Aerial Ladder Truck.
13. Raise, extend, and bed aerial ladder.
14. Name the equipment located in each compartment of the particular apparatus that the probationary Firefighter is assigned at the time of exam.
15. Perform one man salvage cover throw.
16. Perform one man CPR.
17. Place M.A.S.T. on a patient. (in the proper sequence)
18. Place the Hare Traction Splint on a patient. (in the proper sequence)
19. Operate the ambulance gurney, including raising, lowering, chair operations, removing, and placing into the ambulance, etc.
20. Operate the resuscitator, including the use of a mask and nasal canula along with the changing of bottles.

FIRE DEPARTMENT
STANDARDS OF PERFORMANCE
(for rating probationary Firefighters)

1. Station work
   A. Observance of working hours: applies not only to punctuality in reporting for work, but applies to punctuality and readiness at drill time, posted work schedule, meeting schedule, etc.
   B. Cooperation - Team work: Firefighter completes own tasks and helps others in their assignments as needed. Firefighter gets along with others in station. Firefighter is obedient to superiors. Firefighter is considerate to others and cooperates between shifts.
   C. Maintenance of quarters: Firefighter takes responsibility for assigned areas. Cleans according to instruction of Captain. Does not leave other Firefighter's areas dirty. Observes station procedures. Keeps station premises neat, clean and in good order at all times.
   D. Maintenance of apparatus and equipment:
      checks accessories daily for placement and proper operation. Notes repairs needed and reports unusual conditions. Engineer checks apparatus completely each day and reports condition to Captain.
   E. Observance of safety procedures: Firefighter is alert to hazardous conditions. Takes precautions to prevent accidents to self and others. Does not create hazardous conditions or fail to correct them upon observance.
   F. Maintenance of reports and records: keeps reports and records current and according to departmental regulation. Makes legible entries in records. Leaves file in correct order and in good condition. Familiar with reports/records for which responsible.

2. Drills
   A. Knowledge of basic drills: Firefighter demonstrates his knowledge of standard techniques for carrying out basic drills.
      Firefighter improves knowledge of standard procedures when need for improvement is brought to his attention. Continually works toward improving knowledge of standard procedures for basic drills.
   B. Application of standard techniques: follow standard techniques in execution of basic drills. Improves his skill in applying standard techniques when need for improvement is explained. Continually works to improve skill in the application of standard techniques.
   C. Care and use of tools and appliances: know the location and the appropriate use of all tools. Is skillful in the manipulation and use of tools of the job. Cares for and maintains tools and equipment in conformance with manufacturer's specifications and departmental policy.
   D. Mental alertness: Firefighter is aware of the position and task of every member of the team. Observes and anticipate changing conditions affecting the team. Is attentive and receptive to orders an instructions in drills.
   E. Cooperation - team work: Does not resist drills or drilling. Completes his own tasks and helps others in their assignments as needed. Shares his own knowledge and experience with the team. Is receptive to changes in drill techniques.
   F. Observance of safety principles: takes precautions to prevent accidents to self or others. Is alert to hazardous conditions.
      Uses safety equipment and clothing as prescribed by Department.
   G. Knowledge of automotive equipment: has knowledge of and the ability to operate automotive equipment to which assigned.

3. Emergency Work
    A. Adjustment to situation: Firefighter recognizes emergency problems. Evaluates possible alternative actions. Exercises initiative where changing conditions or new conditions make this necessary. Is receptive to changes in strategy.
    B. Response to orders: executes orders promptly. Keeps officers informed of new or changing conditions.
    C. Application of standard techniques: skillful in using standard techniques in meeting emergency situations. Improves skill in applying standard procedures when need for improvement is explained. Continually works to improve skill in the application of standard techniques. Completes basic task within a reasonable time frame.
    D. Cooperation - team work: Carries out own task and helps others in their assignments as needed. Shares job knowledge and experience with others. Adjust to the total job whether shorthanded or not.
    E. Observance of safety principles: takes precautions to prevent accidents to self or others. Is alert to hazardous conditions.
    Uses safety equipment and clothing as prescribed by Department.

4. Fire Prevention
    A. Application of codes: Firefighter follows standard procedure for inspecting an occupancy. Demonstrates ability to recognize hazards.
    B. Thoroughness of inspection: is familiar with those portions of the codes applicable to Fire Prevention. Interprets and applies codes correctly. Keeps reports up to date according to the Departments requirements.
    C. Preparation of reports: maintains files in proper order and in good condition. Is familiar with all reports and records for which responsible.
    D. Public education: establishes a business like relationship with public. Explains code provisions and purposes effectively. Obtains cooperation and understanding of the importance of fire prevention.
    E. Follow-up of inspection: complies with follow-up procedures. Makes follow-up inspections as scheduled.

5. Public Relations
    A. General conduct: Firefighter conducts self, on or off duty, in such a manner as to reflect credit to the Fire service.
    B. Personal appearance: wears proper uniform in the accepted manner. Uniform is neat and clean. Firefighter keeps own person clean and well groomed.
    C. Meeting and handling the public: courteous to the public in all contacts. Gives reliable and accurate information or assistance to the public. Observes Departmental procedures in using the phone.

END OF PROBATION

At the end of your PROBATION PERIOD you will be given a the final exam to see if you have retained the skills and knowledge that you should have gained during the many hours of training that you will have received.

The PROBATION PERIOD will be a minimum of SIX MONTHS, but will vary from one Fire Department to another. At the end of this period you will be given a performance evaluation

At the completion of your PROBATION PERIOD you will be considered a permanent member of the Fire Department which affords job security by means of "due process of the law".

FIRE DEPARTMENT
REPORT OF PERFORMANCE EVALUATION

Name _____ Date _____

Position _____ Period from _____ to _____

Rate each factor (#1-6) as: outstanding; competent; improvement needed; or unsatisfactory.

Check items: + = Strong; 0 = Standard; - = Weak

1. STATION WORK   FACTOR RATING _____
   _____ observance of working hours.
   _____ cooperation - team work.
   _____ maintenance of quarters.
   _____ maintenance of apparatus and equipment.
   _____ observance of safety procedures.
   _____ maintenance of reports and records.

2. DRILLS   FACTOR RATING _____
   House   Apparatus
   _____   _____ Knowledge of basic drills.
   _____   _____ Application of standard techniques.
   _____   _____ Care & use of tools & appliances.
   _____   _____ Mental alertness.
   _____   _____ Cooperation - team work.
   _____   _____ Observation of safety principles.
   _____   _____ Knowledge of automotive equipment.

3. EMERGENCY WORK   FACTOR RATING _____
   _____ Adjustment to situation.
   _____ Response to orders.
   _____ Application to standard techniques.
   _____ Cooperation - team work.
   _____ Observance of safety principles.

4. FIRE PREVENTION   FACTOR RATING _____
   _____ Application of codes.
   _____ Thoroughness of inspection.
   _____ Preparation of reports.
   _____ Public education.
   _____ Follow-up of inspections.

5. PUBLIC RELATIONS   FACTOR RATING _____
   _____ Personal appearance.
   _____ Meeting and handling of public.
   _____ General conduct.

****
COMMENTS: (describe Firefighters strengths and weaknesses. Give examples of work well done and plans for improving performance. Attach additional sheet if necessary)

_____
_____
_____
_____
_____
_____
_____
_____
_____
_____
_____
_____
_____
_____
_____
_____
_____
_____
_____
_____
_____

****

OVERALL EVALUATION
_____ Unsatisfactory.
_____ Improvement needed.
_____ Competent.
_____ Outstanding.

****

SIGNATURES OF REPORTING OFFICERS

This report is based on my observation and/or knowledge. It represents my best judgement of the Firefighters performance.

Rater_____ Date_____

I have reviewed this report.

Reviewer_____ Date_____

Reviewer_____ Date_____

I concur in and approve this report.

Fire Chief_____ Date_____

****

Copy of this report given to Firefighter. Date_____

Report discussed with Firefighter.   Date_____

By:_____ Date_____

****

This report has been discussed with me.

Firefighters signature_____ Date_____

****

Obtain a copy of **"THE COMPLETE FIREFIGHTER CANDIDATE"** for a comprehensive guide on how to become a Firefighter. Available from: INFORMATION GUIDES, P.O. BOX 531, HERMOSA BEACH, CA 90254. 1-800-"FIRE-BKS"

CHAPTER 2

CAREER OPPORTUNITIES AND THE APPLICATION PROCESS

CAREER OPPORTUNITIES AND THE APPLICATION PROCESS

INTRODUCTION:

This chapter will identify the various career opportunities available in the Fire service. As these opportunities are discussed, it will become apparent that obtaining a career in the Fire Service is not an easy task. Much effort must be made by those interested in a Fire Service career. Starting with developing good study habits, identifying the job requirements, the necessary education and background.

Once the information has been obtained, the next part of the process is to properly fill out the job application, resume and cover letter.

This chapter we will discuss all of these topics. It is necessary to understand all these requirements to complete the first step in obtaining a career in the Fire Service.

INFORMATION SHEET #5

TOPIC: CAREER OPPORTUNITIES AND THE APPLICATION PROCESS

INTRODUCTION:
There are many things that you as a candidate should do in your preparation for a career as a Firefighter, the following list will help you to realize what is needed to become a Firefighter.

COMMITMENT LIST:

THINGS TO ESTABLISH
Initiate and establish the following:

1. Good study habits.
2. Good work habits.
3. Your own personal Fire Service Library.

THINGS TO LEARN
Take the time to learn:

1. Job requirements.
2. Information on job flyer.
3. How to file for exams.
4. How to relate in Fire Service situations.
5. How to relate with others.
6. How to apply knowledge that you learn.
7. How to be flexible.
8. How to be a "team player".
9. Various techniques of learning.
10. To read instructions.
11. To follow directions.
12. Fire Service chain of command.
13. Your personal weaknesses.
14. Your personal strong points.
15. How to fill out your application.
16. Duties and responsibilities of a Firefighter.
17. Rescue techniques.
18. Firefighting techniques.
19. Fire Prevention techniques.
20. Manual skills.
21. Foreign language, such as spanish, ETC.
22. As much about the Fire Service as you can.

EXPERIENCE
Acquire experience in:

1. Fire Service written exams.
2. Fire Service oral exams.
3. Fire Service "Cadet" program or Volunteer F.S.
4. Fire station atmosphere. (visit fire stations)

EDUCATION/TRAINING

1. Fire Science courses:
2. Fire Service orientation.
3. Fire apparatus and procedures.
4. Fire hydraulics.
5. Building construction.
6. First aid and CPR.
7. Rescue procedures.
8. E.M.T. 1
9. Fire prevention techniques.
10. Fire Service periodical subscriptions.
11. Fire Service Academy.
12. State Fire Service courses: (usually taken by people already employed in the Fire Service)
    a. Fire Administration, Management/Supervision.
    b. Fire Prevention.
    c. Arson Investigation.
    d. Fire Command.
    e. Instructor courses. ETC!

SELF ASSESSMENT:

NOW is the time to completely evaluate all of the REQUIREMENTS, QUALIFICATIONS, and SPECIFICATIONS for the position of ENTRANCE FIREFIGHTER.

YOU must learn the ones that YOU satisfy and the ones that YOU do not satisfy.

NOW is also the time for YOU to answer any questions that YOU may have about your physical/medical condition.

NOW is the time for YOU to ascertain how much you are willing to commit YOURSELF in order to procure YOUR Fire Service Career! ARE YOU WILLING TO PAY THE PRICE TO BECOME A FIREFIGHTER? if the answer is YES, that means that YOU are willing to do anything to secure a position in the Fire Service.

YOU will now have to take whatever action is necessary to meet any physical/medical requirements that YOU do not satisfy.

YOU now must start to fulfill any and all REQUIREMENTS, QUALIFICATION, and SPECIFICATIONS for the position of ENTRANCE FIREFIGHTER that you do not meet.

YOU MUST BE WILLING TO PAY THE PRICE FOR SUCCESS!

FIRE DEPARTMENT FAMILIARIZATION

A very significant procedure to use in the groundwork for preparing yourself to take an ENTRANCE FIREFIGHTERS exam is to acquire as much information concerning the Fire Department that you are interested in.:

RESEARCH THE MUNICIPALITY, CITY, COUNTY, DISTRICT, ETC.:
Visit the city, county or business administration and collect as much information that you possibly can about the area that you will be serving. Good sources of information are the Chamber of Commerce and the Visitors Bureau.

In order for you to become the best candidate for the position you will need to have as much information as possible.

Some obvious things to learn are, what type of government that the city has, size, population, ETC.

The bottom line is that if you want the position you will spend time to learn:

1. Fire Station addresses and cross streets.
2. Where particular apparatus are located in the City, what apparatus each station houses.
3. The Departments chain of command.
4. The cities government.
5. The condition of the City water system.
6. The type of fire hydrants in City.
7. Water supply, source, grid system, etc.

If another candidate knows all of the above you want to know all of the above plus more!

In reality all of this research does not take that much time, so take the time and do it right.

VISIT THE FIRE STATIONS:

Visiting Fire Stations can be one of the most informative steps that you can take in researching a Fire Department.

Don't just visit the Headquarters station, try to visit all of the stations in the Department.

Visiting different stations allows you to become familiar with various functions that a station may have, such as:

1. Airport station.
2. Wild-land area.
3. High-Rise area.
4. Hazardous Materials station.
5. Harbor area. ETC!

Visiting Fire Stations will allow you:

1. To meet people that may be on your interview board.
2. To project a feeling to the Department that you are interested enough in the Department to research it.
3. To be more prepared to answer questions about the Department.
4. Prepare you for some obvious questions such as:

    Can you tell us what have you done to prepare yourself for a job as Firefighter with this Department?

    What can you tell us about this Department?

    Why did you select this Fire Department for your career in the Fire Service?

When you visit these Fire Stations take a list of questions with you and ask them!

It is beneficial to talk to new Rookie Firefighters that have just recently been through what you are going through. It is also beneficial to talk with experienced personnel since they will know current and traditional information concerning the Department.

Some of the questions that you should ask are:

1. What kind of schedule does this Department work? hours, etc.
2. When does the probationary period begin and how long does it last?
3. What will the Department expect of me during the probationary duration?
4. Does this Department have more than one Fire Station? if yes how many?
5. Does each station have a specific function? if yes what?
6. What are the manpower levels of each station, and the Department?
7. Can you tell me anything about the testing process and the hiring board?
8. Are you aware as to what the Department is looking for in each candidate?
9. Find out if it is possible for you to participate in a ride along program.
10. Ask if you can look at the apparatus and equipment in the station, including any information that might be pertinent for the exam.
11. Find out if they have any study material concerning the way they drill, operations manuals, and yearly annuals that you might be able to look over.

EXAM PREPARATION

KNOW WHO IS GETTING HIRED/KNOW THE COMPETITION :

Spend time talking with other prospective Firefighter Candidates and recently hired Firefighters, try to find out what they have been doing or what they have done to capture the position of Firefighter. This is an excellent way to find out if you are forgetting to cover all of the bases.

ESTABLISH AN EXAM CHECK LIST :

1. Communicate with Personnel Department.
2. Communicate with Fire Department.
3. Get advice from Fire Science instructors.
4. Investigate for job openings in newspaper.
5. Acquire job specifications.
6. Acquire the exam announcement.
7. Obtain the jobs medical standards.
8. Get the job application.
9. Complete the job application.
10. Formulate a resume.
11. Submit the job application.
12. Submit the resume and develop a "Cover Letter".
13. Prepare for written exam.
14. Prepare for physical agility exam.
15. Prepare for oral exam.
16. Learn location of all portions of exam.
17. Make visits to the Fire Stations.
18. Research particulars of the Fire Department.
19. Research the particulars of the community.
20. Take practice oral exams.
21. Select clothing for the oral interview.
22. Read fire Service related books and magazines.
23. Acquire a Fire Technology college degree.
24. Get involved in civil service activities.
25. Stay healthy:
26. Proper diet.
27. Exercise.

PRACTICE EXAMS :

The only way to get proficient at taking Fire Service Exams is to take practice exams whether it be with work books, at school, from Fire Department friends and acquaintances, or probably the best way is to take any and all actual Fire Department exams that you possibly can. Although you should be aware that many Fire Department frown on the exercise of taking their exams just for practice!

Each time you take an actual exam you will gain priceless test taking experience and procedures that cannot be achieved in any other format.

You will discover:

1. The feeling of actual test situations.
2. How to perform in a stressful atmosphere.
3. How to assess your own abilities, knowledge, and progress.

These discoveries will allow you to adjust and take instant and appropriate actions to improve wherever needed. This will make you elevate your preparedness for future exams. Regardless of how much reference material that you cover, there will always be the obscure and unanticipated obstacles that will come up. Obviously the more exams that you participate in the more opportunity you will have to be exposed to these obstacles.

When you have taken a few exams you will begin to discover that many of the written exams will have the same type of questions or even the exact same questions will appear on different exams.

On occasion you may encounter the same exact written exam that you may have previously taken.

You will also come to discover that you will encounter similar events in the physical agility portion of the exam. Many Fire Departments use similar physical ability examinations.

On the oral interview portion you will find the same type of questions are being asked, although they may be presented differently.

IN ALL PORTIONS OF THE EXAMS THERE WILL BE A SPECIFIC PATTERN AND PROCESS THAT WILL APPEAR!

If you choose to take exams with Fire Departments that you don't necessarily wish to work for because they do not meet all of your pre-conceived criteria, you are not obligated to accept a position if it is offered to you. Just be aware that the job with all the criteria that you want, may never become available to you.

LOCATING EXAMS

Some of the ways that you can locate exams are:
1. Contacting various area Personnel Departments.
2. Directly contacting various Fire Departments.
3. Through the Fire Technology Coordinator or the Fire Technology Department of your local Community College and in the Fire Technology courses.
4. In newspapers.
5. Firefighter Information Services/Recruitment Services.
6. Fire Service Directories.

PERSONNEL DEPARTMENTS:

You may find out about exams through the Personnel Department of many areas by directly visiting, telephoning, or by writing them. It is usually best to go in person if at all feasible. Contact as many jurisdictions as reasonably can!

Information to accumulate should include:
1. The status of current hiring list and how long that it is certified.
2. When their last exam took place.
4. How many Firefighters were hired from the current list.
5. When will the next Firefighters exam take place.
6. How many Firefighter openings will be available.
7. If there are any special requirements.
8. Do they have a affirmative action policy.
9. Do they have a residence requirement.
10. Are there any language requirements.
11. Acquiring the last and/or current job flyer announcement for Firefighter.
12. Will applications be limited.
13. When and where will they give out applications and will there be a time limit.
14. Will they notify you of the exams and when to apply.
15. Can you leave your name, address, and telephone number.

DON'T BE SHY, THEY WILL ASSIST YOU! THEIR OBJECTIVE IS TO RECRUIT THE BEST POSSIBLE CANDIDATES THAT THEY POSSIBLY CAN RECRUIT.

FIRE DEPARTMENTS:

Many Fire Departments will have their own Personnel Director. You should direct the previously mentioned questions to this person.

In addition the Fire Department Personnel Director may be able to inform you as to any personnel on the Department that may be anticipating or even contemplating retirement.

The Fire Department Personal Director is likely to be aware of any possible additional Fire Companies that may be going into service and of new/additional Fire Stations that may be opening, or any other reason that staffing may be increasing on the Department.

Another benefit in going directly to the Fire Department is that they are usually kept up-to-date of what is going on in the Fire Service including exams in other jurisdictions. You may observe exam flyers that are posted ion the Fire Station bulletin board.

You may get someone such as a Rookie Firefighter or a interested member of the Department to notify you when they become aware of Fire Department exams.

DON'T BE AFRAID TO ASK! FIREFIGHTERS IN GENERAL ENJOY HELPING OTHERS.

COMMUNITY COLLEGE FIRE TECHNOLOGY DEPARTMENTS:

Many Community Colleges now offer Fire Technology Curriculums. These Colleges are usually affiliated with State Fire Training Departments and the State Department of Education.

The Fire Technology Curriculums are usually administered by a Fire Technology Coordinator that is usually also employed by a Fire Department in a nearby jurisdiction.

Most Fire Technology Curriculums will have an Introduction to Fire Science course, which is a type of pre-employment course of the Fire service.

Go to your local Community College and set up an appointment with the Fire Technology Coordinator. In your meeting with the Fire Technology Coordinator ask as many of the previously mentioned questions that are applicable. Fire Technology Coordinators are usually on a mailing list to receive Fire Department announcements for exams. Most will keep a record of the announcements and will be glad to assist you.

While visiting the Community College, enroll in various Fire Technology courses if you have not done so already. Aside from the obvious benefits of these classes you can always find out about Fire Service exams from other students. Many of the students are already Firefighters and others are in the same boat as you, in either case they are going to be aware of upcoming exams. MAKE IT A POINT TO TALK WITH OTHER STUDENTS!

PERIODICALS/NEWSPAPERS:

Many Fire Departments will advertise in periodicals/newspapers in their particular areas. The advertisement may appear as a "space ad" in the selected sections, such as the sports section, or any other section of the newspaper. The advertisement may also be in the form of a classified ad located in the want adds of periodicals/newspapers.

Some of the larger Fire Departments will place adds in newspapers of various newspapers in order to attract some of the high caliber candidates from other areas.

A good place to look for these adds is in the newspaper racks of the main Library in your area or maybe even in your local Library. BE SURE TO SPEND THE TIME TO CHECK!

FIREFIGHTERS INFORMATION SERVICES:

There are Firefighter Information Services that are computerized to link Entrance Firefighter Candidates with all of the available Fire Departments that are testing.

Firefighter Information Services will charge a fee to save you some legwork and notify you when Fire Departments are accepting application for test.

Firefighter Information Services will send you computer print-outs whenever a Fire Department is going to accept applications. This print out may include a district map which will pin-point where the test is going to be held.

Firefighter Information Services may also provide information regarding such details as:
1. Square miles of jurisdiction testing.
2. Population of the area that is testing.
3. Number of engine companies that are maned.
4. Staffing of Fire Department that is testing.
5. The number of the Departments Fire Stations.
6. The number of responses that the Fire Department makes annually.

One example of a Fire Information Service is **PERFECT FIREFIGHTER CANDIDATE**: which at this time covers several western states and will continue to add other states in the future. On the following page are two examples of the information post card that **PERFECT FIREFIGHTER CANDIDATE** sends out to prospective Firefighters that subscribe:

FOLSOM is accepting applications for: PARAMEDIC Until: May 31, 1991 Salary: $2,183. per mo. Requires: Two years exp. as a EMT- II or Paramedic, Calif. Paramedic Cert. For more info contact: City of Folsom, Personnel Dept., 50 Natoma St., Folsom, Ca. 95630 (916) 355-7200

LYNWOOD is accepting applications for: FIREFIGHTER TRAINEE Until: June 6, 1991 Salary: $4.77 per hr. Requires: 18 Yrs old, VCDL, Reside within 20 min. of City of Lynwood, HS/GED, . For more info contact: City of Lynwood, Personnel Office, 11330 Bullis Rd., Lynwood, Ca. 90262 (213) 603-0220

CAMPBELL is accepting resumes for: FIRE CHIEF Until: May 31, 1991 Salary: $72,948. - $87,540. per yr. Requires: Experienced professional administrator. For more info: Hughes, Heiss & Assoc., 675 Mariners Island Bl. #108, San Mateo, Ca. 94404 (415) 570-6111

SAN MATEO AGENCIES JOINT RECRUITMENT is accepting applications for: FIREFIGHTER Until: ONE DAY ONLY June 15, 1991 9:00 a.m. - 1:00 p.m. @ Main Cafeteria, San Mateo College, 1700 W. Hillsdale Bl., San Mateo, Ca. To register you must present your valid Calif. Drivers Lic. Salary:Avg. Starting @ $2,500. per mo. Requires: 18 Yrs old, HS/GED, VCDL. Nine Depts. are participating in this recruitment (Burlingame, Foster City, Hillsborough, Millbrae, Redwood City, Half Moon Bay, Menlo Park, Woodside, & South County.) For more info call: (415) 361-0994

ALBANY is accepting applications for:
FIREFIGHTER/PARAMEDIC Until: June 4, 1991 Salary: $2,862 - $3,483. per mo.Requires: 18 Yrs old, VCDL, HS/GED, Paramedic Cert. accreditation in Alameda Co. or any nine Bay area counties. For more info contact: City of Albany, 1000 San Pablo Ave., Albany, Ca. 94706 (415) 528-5710

LA HABRA HTS. is accepting applications for: VOLUNTEER-FIREFIGHTER Until: 8:00 a.m. June 1, 1991 Requires: 18 Yrs old, HS/GED. Applications must be picked up in person @ La Habra Hts Fire Dept., 1245 N. Hacienda, La Habra Hts., Ca. 90631

FOLSOM (Sacramento Co.)
ALBANY (Alameda Co.)
CAMPBELL (Santa Clara Co.)
JOINT RECRUITMENT (San Mateo Co.)
LYNWOOD
LA HABRA HTS. (Los Angeles Co.)
District #1

ADDRESS OF THIS SERVICE:
Perfect Firefighter Candidate
P.O. Box 7046
Oxnard, CA 93031
Phone: (805) 984-7065 and (805) 658-7065 or TOLL FREE: 1-800-326-8401

The following is a list of a few of the other Firefighter Information Services that are available:

Emergency Concepts
P.O. Box 1402
Sun Valley, CA 91353-1402

Sunbelt Opportunities
P.O. Box 16578
Chapel Hill, NC 27516

National Emergency Services Trade Employment Guide
(N.E.S.T.E.G., INC.)
P.O. Box 176
Queens, NY 11419; Phone: (718) 738-6315

Professional Fire Service Recruitment
Box 7254
Warwick, RI 02887

Western States Employment Journal
P.O. Box 3156
Costa Mesa, CA 92628

Fire Search
7422 Anchorage
Hilton Head Island, SC 29928

Rescue
P.O. Box 2992
Winchester, VA 22601

Rocky Mountain Employment Newsletter
703 S. Broadway, #100
Denver, CO 80209; Phone: (303) 988-6707

LENCO
1125 B, Arnold Drive, Suite 240
Martinez, CA 94553; Phone: (415) 746-7085

FIRE SERVICE DIRECTORIES:

Obtain a Fire Service Directory from your States Fire Marshal's Office. There may be a charge for this directory.

A Fire Service Directory will have a list of all the Fire Departments in your state. The Fire Service Directory will include contact names and telephone numbers for your use. There will be other pertinent information in these directories.

Use the telephone numbers in the Fire Service Directory to call and see if you can leave your name, address, and telephone number on a waiting list so that you can be notified as to when each particular Fire Department will be accepting applications. Many Fire Departments or their Personnel Departments will accept "interest cards" and mail them out to individuals when applications are going to be accepted. These cards usually have an expiration date, be sure to record this date so that you will not forget to renew it.

FIREFIGHTER JOB ANNOUNCEMENTS

EXAMPLE #1

CITY OF _____

INVITES APPLICATIONS FOR

FIRE FIGHTER
OPEN

SALARY:	$2445 to $3120 per month plus excellent benefits.
APPLICATION PROCESS:	Application packets and instructions will be available through September 22, 1989 from the Civil Service Department between the hours of 8:00 AM and 5:00 PM. You may request an application packet by mail by phoning 618-2969.
	APPLICATIONS MUST BE RETURNED BY U.S. MAIL ONLY! WALK-IN APPLICATIONS WILL NOT BE ACCEPTED.
	ONLY APPLICATIONS POSTMARKED ON MONDAY, SEPTEMBER 25, 1989 OR TUESDAY, SEPTEMBER 26, 1989 WILL BE ACCEPTED. 500 APPLICATIONS WILL BE RANDOMLY SELECTED FROM THOSE ENVELOPES WITH THE PROPER POSTMARK. NO EXTENSIONS WILL BE ALLOWED.
REQUIREMENTS:	**Smoking** - Must be a non-smoker. Smoking on or off duty is prohibited.
	E.M.T. - New employees must obtain a _____ Emergency Medical Technician I Certificate prior to the completion of the first year of employment.
	Education - Equivalent to graduation from high school.
	License - Valid driver's license.
	Age - At least 21 years of age by final filing date.
	Vision - 20/40 uncorrected with both eyes (the weakest 20/70; the stronger 20/40). Each eye corrected to 20/30.
	Hearing - Not more than 10% loss in either ear. Detailed medical standards are on file in the Personnel Office.
SELECTION PROCEDURE:	**Written** - Designed to measure a candidate's ability to understand and follow oral and written instructions; understand diagrams, tools, equipment and three-dimensional objects; perform math calculations; and ability to successfully interact in a community living situation.
	Interview - Designed to measure a candidate's orientation to a career in fire service; motivation; maturity; oral communication skills; and interpersonal relations.
	Scoring - The final overall score will be a combination of the written test score (weighted 50%) and the interview score (weighted 50%). Prior to selection, candidates must qualify on a physical agility test (pass/fail only), background investigation; medical examination; and psychological examination. Information on the physical agility test may be found at any Torrance Library reference desk.
TEST DATES:	The written test is tentatively scheduled for Monday, October 16, 1989 with interviews tentatively scheduled for Tuesday, October 31 and Wednesday, November 1, 1989.

SEE APPLICANT INFORMATION ON REVERSE SIDE
EQUAL OPPORTUNITY/AFFIRMATIVE ACTION EMPLOYER

REVERSE SIDE OF EXAMPLE #1

APPLICANT INFORMATION

APPLICATION
Applications must be received by the final filing date (postmarks are not accepted) unless otherwise authorized by the Civil Service Department. Applications can be accepted only if they clearly indicate that the minimum requirements have been met by the final filing date. The experience requirement refers to experience which is full time, paid working experience. Part-time and volunteer experience will be credited on a proportionate basis.

All applications accepted at time of filing are subject to further review and verification and may be rejected at any time if it develops that requirements for admission to the examination are not met.

Admission to an examination does not insure that a passing score will be received on any part of the examination.

VETERAN'S PREFERENCE
Veteran's preference is given for service during certain periods of war or national emergency on open Civil Service examinations only, and may be claimed by presenting proof (e.g. DD214) of honorable discharge or release at time of filing the application. It is the applicant's responsibility to present the evidence when submitting the application.

EXAMINATION
Applicants for Civil Service positions are required to take a competitive examination, which may include written tests, performance tests or an oral interview.

A passing grade must be obtained on each part before applicants are permitted to take succeeding parts. A written test, when not standardized or copyrighted, will be available for review for five (5) working days immediately following its administration.

HANDICAPPED
A handicapped applicant is defined as any person who has a physical or mental impairment which substantially limits one or more major life activities or whose employment is negatively affected by decisions based on a record or perception of such impairment.

The Civil Service Department will attempt to provide special testing arrangements for handicapped applicants. If you feel you have a need for special testing arrangements you must notify the Civil Service Department prior to the final filing date.

APPOINTMENT
It is the applicant's responsibility to notify the Civil Service Department of any change of conditions affecting employment, such as applicant's change of address or phone, or willingness to accept temporary employment, or similar types of information.

Before appointment to a position, successful applicants must pass a medical examination and, if required, a psychological examination. Detailed medical standards are available in the Personnel Department.

A background check will be conducted on all candidates for City positions. Fingerprints will be checked.

BENEFITS
The City offers a wide variety of benefits, which vary by employee classification. All employees are granted excellent vacation, holiday, and sick leave benefits; with sufficient years of service employees may earn up to 27 days per year of vacation and accrued sick leave may be redeemed in cash. The City provides up to 12 months paid industrial injury leave; a long-term disability plan to cover non-job related illnesses or injury; comprehensive family medical coverage and eligibility for dental and vision plans and life insurance; credit union membership; eligibility to borrow up to 40 hours of vacation; eligibility to participate in a deferred compensation plan; and membership in the State Public Employees Retirement system (most employees are also covered by Social Security). Some employees are also eligible for up to 10% longevity pay; educational incentive pay; premium pay for special assignments; tool and safety shoe allowance; uniform allowance or City provided uniforms; periodic medical examination and City payment of the employee's share of retirement costs. Information on the specific benefits granted each position may be obtained from the Personnel Department.

CITY OF
Civil Service Department

EXAMINATION
ANNOUNCEMENT

**FIREFIGHTER JOB ANNOUNCEMENT
EXAMPLE #2**

CIVIL SERVICE EXAMINATION for
FIREFIGHTER

Firefighter's are Salaried Employees.
The Approximate Average Annual Salary for Entry-Level Firefighters is $ _____

Closing Date: _____ Examination Date: _____
Deadline for Submitting application By Appointment Only

Examination for Firefighter will not be offered again until the annual test period currently scheduled to begin March 1990. Vacancies which occur prior to availability of that eligibility register may be filled from the existing eligibility register.

REQUIREMENTS

AGE: Applicant must have reached his/her 18th birthday by the above stated deadline for submitting an application. A birth certificate must be submitted with the application; however, the deadline for submitting the birth certificate is _____

EDUCATION: Be a High School graduate or possess equivalency. A diploma or certificate must be submitted with the application; however, the deadline for submitting the diploma or certificate is _____

DRIVER'S LICENSE: All applicants are required to possess or obtain a _____ Driver's License.

SELECTION PROCEDURES

All applicants must successfully complete the written test and physical performance test, prior to placement on the eligible list.

SUBJECTS ON WRITTEN EXAM: Following instructions, reading comprehension, depth perception, mechanical aptitude, math, emergencies, and effective working relations.

Prior to final selection and appointment, all applicants must submit to, and be approved by the following procedures:

— — Background Investigation
— — Polygraph Examination
— — Extensive Physical Examination including Drug Testing and weight in proportion to height*
— — Comprehensive Psychological Evaluation

*A copy of the height and weight chart is available upon request. Questions should be referred to the Civil Service Office.

BENEFITS

Liberal leave policy: 10 days vacation per year with less than 5 years of service; 15 vacation days per year with more than 5 but less than 10 years of service; 20 days vacation per year with more than 10 years of service. Sick leave — 12 days per year accumulative. 9 paid holidays; attractive pension plan; may retire after reaching 50 years of age and 25 years of service; Uniform allowance of $500 initially and $300 each year thereafter; job security; longevity pay increases. Insurance — Hospitalization Insurance or Health Maintenance Organization and $15,000 life insurance plus $25,000 one time death benefit in line of duty. In addition, the City provides a Wellness Program and Employee Assistance Program.

_____ IS AN EQUAL OPPORTUNITY EMPLOYER

GENERAL INFORMATION CONCERNING CIVIL SERVICE EXAMINATIONS

APPLICATION:
All applicants must apply through the Civil Service Merit Board Office.

CITIZENSHIP:
Must be a citizen of the United States.

VETERANS PREFERENCE:
Veterans preference credits will be allowed in accordance with Civil Service Rules, providing a copy of DD-214 is filed. Disability if claimed must be supported by proof, plus amount and must be filed before date of examination.

APPOINTMENTS:
As vacancies occur they will be filled from the Eligible Register in accordance with Civil Service Rules and Regulations.

MEDICAL EXAMINATIONS:
All Civil Service employees must complete a comprehensive physical examination prior to employment in accordance with Civil Service Merit Board rules.

AGE:
Applicants must be not less than the minimum and must not have reached the maximum age stated above (if there are age restrictions).

EXAMINATION:
Examinations will be job related.

EXAMINATION RESULTS:
You will be notified by mail of the results of the examination.

CIVIL SERVICE MERIT BOARD

JOB APPLICATIONS

When you have decided to apply for a Firefighters position with a particular Fire Department that will be administrating an exam, the first thing that you will do, as an applicant, is acquire an APPLICATION.

Each Fire Department will have its own way of distributing their application. The number of applications handed out is usually restricted to a specific number.

Some Fire Departments may distribute the applications out over a specified time period. (usually limited)

Some Fire Departments may post a time when you may begin to line up to receive one of a limited number of applications that they are going to distribute.
(it is very important to show up early for this type of job application distribution)

Other Fire Departments may use a lottery system to determine which job applicants will be allowed to receive an application.

Many Fire Departments will distribute an unlimited number of applications and then accept only the ones that they have deemed to have contained all the necessary requirements and qualifications for the position.

RULE #1: FOLLOW BOTH VERBAL AND WRITTEN DIRECTIONS!

Prior to filling out your application make a photo copy to use as a rough draft.

When you complete the rough draft go over it completely and carefully, then have it proofread by someone that is adapt with spelling, grammar, punctuation, etc.

If it is possible applicants should have someone affiliated with the Fire Service, such as a Fire Technology Coordinator and/or Instructor review the application.

Always use a typewriter to complete your application form. BE NEAT!

Make sure that you sign and date the application.

When answering questions concerning the leaving of prior employment, use a positive reply such as: "advanced to a more responsible capacity" or "moved to continue/complete/advance education"

Make sure that you complete all the spaces on the application. If a space does not pertain to you, use the designation N/A. (not applicable)

Be prepared to have documents that may be required with the application.

Documents that are usually required are those that will verify the requirements and qualifications that appear on the job announcement.

Examples of documents that are usually required for verification are:
1. Various classes of drivers licenses.
2. Military discharge documents.
3. College/University degrees or certificates.
4. Special training/courses certificates.
5. ETC.

Be aware that the application is the only connection that you will have with the oral interview board members.

Don't forget to make a copy of your application.
There are two reasons for making a copy of your application:
1. You will want to study the application and memorize it for your interview.
2. You can also use it as a master for future job applications.

TURN IN YOUR APPLICATION PRIOR TO THE CLOSING DATE!

The following pages contain examples of how various jurisdictions job applications look and what information they require.

JOB APPLICATION
EXAMPLE #1

EQUAL OPPORTUNITY/AFFIRMATIVE ACTION EMPLOYER
APPLICATION FOR EMPLOYMENT
CITY OF: _____

I. GENERAL INFORMATION *Please type or print in ink*

Position applied for: _____

Name _____
      First   Middle   Last

Home Address _____
              Number   Street   City   State   Zip Code

Telephone _____ Business _____
      Area Code   Area Code

Social Security Number _____/_____/_____ California Driver's License Number _____

When applicable do you meet the age requirements for the position you are applying for and can you provide document of proof? YES [ ] NO [ ]

Upon appointment can you provide proof of legal residency in the U.S.? YES [ ] NO [ ]

II. TRAINING/EDUCATION

Circle Highest Grade Completed   High School Last Attended   Location   Graduate YES [ ] NO [ ]
1 2 3 4 5 6 7 8 9 10 11 12   G.E.D. YES [ ] NO [ ]

College, Business, or Trade School(s) attended and location   Major Subjects   Semester Units   Degree(s)

_____
_____
_____

Please list any technical or professional licenses/credentials you possess that are required by or related to this position. Also, list the expiration dates:
_____
_____
_____

Active Military Duty: From _____ To _____ Highest Rate _____ Branch _____

Please describe any pertinent training you received while in the armed forces: _____
_____
_____

Type of Discharge: _____

III. REFERENCES Give names and addresses of three references other than relatives or former employers.

Name   Address   Phone   Occupation
_____
_____
_____

EQUAL EMPLOYMENT OPPORTUNITY PROGRAM

The following confidential information is requested as part of the Equal Employment Opportunity Program as required by the Civil Rights Act. Your cooperation in this procedure will assist the City in providing equal employment opportunity to all prospective candidates.

Exact Job Title: _____

Race (ethnicity) [ ] White   [ ] Asian/Asian surname
                 [ ] Black   [ ] American Indian
                 [ ] Spanish/Spanish surname [ ] Other _____

[ ] Male   [ ] Female

Where did you hear of this opportunity? Newspaper _____ Radio _____
                                         (name)
City Bulletin _____ Friend _____

REVERSE SIDE OF EXAMPLE #1

IV. PREVIOUS EMPLOYMENT

List most recent experience first. Carefully account for all employment for at least the last ten years. Complete this section and attach supplemental sheets or resumes if you desire.

1. Title _____ From _____ To _____
   Current Employer _____ Supervisor _____
   Address of Employer _____
   Salary _____ Reason for leaving _____
   Duties _____
   _____ May we contact your present employer? YES [ ] NO [ ]

2. Title _____ From _____ To _____
   Employer _____ Supervisor _____
   Address of Employer _____
   Salary _____ Reason for leaving _____
   Duties _____

3. Title _____ From _____ To _____
   Employer _____ Supervisor _____
   Address of Employer _____
   Salary _____ Reason for leaving _____
   Duties _____

4. Title _____ From _____ To _____
   Employer _____ Supervisor _____
   Address of Employer _____
   Salary _____ Reason for leaving _____
   Duties _____

V. SUPPLEMENTARY INFORMATION

1. Are you related to anyone working for the City _____ ?   YES [ ]   NO [ ]
   If yes, what department _____
   What is your relationship? _____

2. Have you ever been convicted or placed on probation for any felony?   YES [ ]   NO [ ]

3. Have you ever been discharged from a position or resigned upon request to avoid discharge?   YES [ ]   NO [ ]

4. Have you ever received worker's compensation for an industrial (on-the-job) injury?   YES [ ]   NO [ ]

5. Do you have any physical handicap, chronic disease, or other disability that would affect your ability to perform in the position applying for?   YES [ ]   NO [ ]

NOTE: An answer of YES to any of the above questions must be accompanied by explanatory remarks in the section provided below or on additional sheets.

Remarks: _____
_____
_____

IF ADDITIONAL SPACE IS NEEDED, PLEASE ATTACH A SEPARATE SHEET

I hereby certify that the answers given by me to the foregoing questions and statements made are true and correct. I understand that knowingly making a false statement or omission on this application may be deemed sufficient cause for rejection of this application or dismissal after employment.

Read the above statement

Sign Application Here _____   Date _____

JOB APPLICATION
EXAMPLE #2

CITY OF _____
EMPLOYMENT APPLICATION

THE CITY OF _____ PROVIDES A PUBLIC PERSONNEL SYSTEM BASED ON MERIT PRINCIPLES. IT SERVES FOR CONSTANT IMPROVEMENT OF PUBLIC SERVICE BY EMPLOYING AND DEVELOPING THE BEST QUALIFIED PEOPLE AVAILABLE. EVERY JOB APPLICANT IS RATED SOLELY ON HIS ABILITY WITHOUT REGARD TO RACE, COLOR, CREED, SEX, POLITICAL BELIEFS. ALL PERSONS EMPLOYED BY THE CITY OF KNOXVILLE MUST BE LEGALLY AUTHORIZED TO WORK IN THE UNITED STATES.

READ THROUGH THE ENTIRE APPLICATION THEN ANSWER EVERY QUESTION IN INK. WRITE NO OR NONE AFTER QUESTIONS THAT DO NOT APPLY TO YOU.

Date _____ Job Position Interested In: _____

Print Full Name _____
First / Middle / Last

Present Address _____
(Number) (Street or Route) (City) (State) (Zip)

Telephone No.
Home: _____
Business: _____
Ext.: _____

Are you now employed? Yes ☐ No ☐ If yes, may we refer to your present employer? Yes ☐ No ☐
Have you ever been discharged or requested to resign from any position? Yes ☐ No ☐

SOCIAL SECURITY NO. _____

If yes, explain: _____

List and describe all your significant work experience. Include experience in military trades or occupations. Be specific. Start with your present or last position and work back.

Have you ever been convicted of a felony, misdemeanor or of any law, ordinance, or police regulation (excluding traffic violations)? Yes ☐ No ☐

If yes, explain fully what, where and results (i.e. paid fine, served time in jail)

Last or Present Job

Employing Firm _____

From ___ Month ___ Year

Firm Address _____

NOTE: The existence of a criminal record does not constitute an automatic bar to employment.

To ___ Month ___ Year

Supervisor's Name _____

Do you have any physical or mental conditions which would preclude your performing the duties of the position for which you are applying? Yes ☐ No ☐
If yes, explain

Hours per week _____

Your Title _____

Starting Salary $ ___ per ___

Specific Duties _____

Last Salary $ ___ per ___

Reason for Leaving _____

If you are applying for a position which requires a driver's license, answer the following:
Do you possess a Driver's License? Yes ☐ No ☐
If yes, what State? _____
Number _____

Employing Firm _____

From ___ Month ___ Year

Firm Address _____

-- VETERANS --

Have you ever served in the Armed Forces or any reserve component of the United States? Yes ☐ No ☐
If yes, submit a copy of your discharge (DD214) with this application. If yes, was your discharge other than "honorable" or "under honorable" conditions? Yes ☐ No ☐ If yes, explain fully:

To ___ Month ___ Year

Supervisor's Name _____

Hours per week _____

Your Title _____

Starting Salary $ ___ per ___

Specific Duties _____

Last Salary $ ___ per ___

Reason for Leaving _____

Date Enlisted	Date of Separation
Rank at Separation	Branch of Service

Employing Firm _____

From ___ Month ___ Year

Firm Address _____

To ___ Month ___ Year

Supervisor's Name _____

Hours per week _____

Your Title _____

Starting Salary $ ___ per ___

Specific Duties _____

Did you receive a medical discharge?
Yes ☐ No ☐

Last Salary $ ___ per ___

If yes, do you have a disability rating?
Yes ☐ No ☐

CSS-100

Reason for Leaving _____

Percent disabled: _____ %

REVERSE SIDE OF EXAMPLE #2

Are you skilled in the operation of any office machines, shop tools, or heavy equipment (such as typewriter, accounting machines, welding equipment, jackhammer, etc.)? Yes ☐ No ☐

If yes, name the type of machines or equipment, and the years of experience you have had in the space provided below:

Are you skilled in any trade? Yes ☐ No ☐

If yes, name:

_____

_____

_____

_____

Elementary and High School Education

Highest grade completed (circle one) 1 2 3 4 5 6 7 8 9 10 11 12 Do not circle 12 unless you have received a diploma from the high school	Date Last Attended (Month & Year)	Name and Location of Last School attended (high school, junior high, or elementary)	Do you possess a certificate of high school equivalency (G.E.D)? ☐ Yes ☐ No

Colleges and Universities Attended

Name & Location	Dates Attended (Mo. & Yr.) From / To	Credit Hrs. Semester / Quarter	Degrees Received

Chief Undergraduate College Subjects	Credit Hrs. Semester / Quarter	Chief Graduate College Subjects	Credit Hrs. Semester / Quarter

Are you now working toward a high school diploma or a college degree (Bachelor's, Graduate, or special)? Yes ☐ No ☐ If yes, name the diploma or degree and give the date you expect to receive it.

Special Training (correspondence, business, trades, vocational, Armed Forces schools, etc.)

Name and location of schools, training centers, institutes, etc.	Dates Attended (Mo. & Yr.) From / To	Courses or Subjects Taken	Certificates Received or other pertinent information

PLEASE READ CAREFULLY BEFORE SIGNING

I hereby certify that I have never been a member of any organization or group which seeks to alter the form of Government of the United States by unconstitutional means. I further certify that all answers to the above questions are true and I understand that any misstatement of material facts contained in this application will cause forfeiture upon my part of all rights to any employment subject to the jurisdiction of the Civil Service Merit Board of the City of _____ for three years.

Do you hold any trade or professional license or permit? Yes ☐ No ☐

If yes, give title or kind in space provided below:

_____

_____

_____

_____

_____ _____
Signature   Date

THIS APPLICATION CANNOT BE PROCESSED WITHOUT A SIGNATURE

AN EQUAL OPPORTUNITY EMPLOYER

SUPPLEMENT TO EXAMPLE #2

CITY OF _____
CIVIL SERVICE DEPARTMENT
SUPPLEMENT TO THE JOB APPLICATION

NAME _____ DATE _____

POSITION APPLIED FOR _____

PART I: REPORT OF CONVICTIONS

In the spaces below, give complete details for every time that you have been convicted of any violation of law. A conviction includes a plea, verdict or finding of guilt, regardless of whether a sentence was imposed by the court. You must include any arrest for which you are now out on bail (or on your own recognizance) pending trial.

Do not include any record of arrest or detention which did not result in a conviction or any record regarding a referral to and participation in any pre-trial or post-trial diversion program. Do not include any conviction for violation of subdivision (b) or (c) of Section 11354 of the Health and Safety Code or a statutory predecessor thereof, or subdivision (c) of Section 11360 of the Health and Safety Code, or Section 11357, 11365 or 11550 of the Health and Safety Code as they relate to marijuana.

A conviction does not automatically mean you cannot be appointed. The nature and gravity of the offense, the length of time since the conviction, any evidence of rehabilitation, and the duties of the position for which you have applied are important. Give all of the facts so that a decision can be made.

Begin with your first conviction. If you need more space, please continue on the back.

WARNING: You may be disqualified or dismissed from employment with the City of _____ unless you fill out Part I of this form *accurately* and *completely* to the best of your knowledge. This form will be immediately separated from your application and will not be considered in the hiring process except as provided by Section _____.

☐ IF NO CONVICTIONS, CHECK BOX, SIGN AND DATE BELOW.

APPROXIMATE DATE, CITY AND STATE OF CONVICTION	OFFENSE FOR WHICH CONVICTED	DESCRIBE BRIEFLY THE CIRCUMSTANCES CONCERNING YOUR CONVICTION INCLUDING AMOUNT OF FINE, LENGTH OF JAIL OR PRISON SENTENCE, AND LENGTH OF PROBATION PERIOD.

I certify that the above is true and correct.

_____   _____
Signature   Date

CIVIL SERVICE DEPARTMENT

- -

PART II: APPLICANT INFORMATION

Your *voluntary* cooperation in completing Part II of this form is appreciated. This information is used for statistical purposes only and will be immediately separated from your application upon receipt. It will have no effect on either your examination or the hiring process. See back side for definitions.

SEX: ☐ Male ☐ Female

ETHNIC CATEGORY:
☐ "Black"
☐ "Hispanic"
☐ "White"
☐ "American Indian or Alaskan Native"
☐ "Asian or Pacific Islander"

HOW DID YOU LEARN ABOUT THIS JOB OPENING?
☐ City employee   ☐ Newspaper
☐ Walk-in   ☐ Other _____

HANDICAPPED: ☐ Yes ☐ No   If yes, please explain: _____
_____
_____

RESUMES

Most applications do not allow appropriate space for emphasizing your positive points or to add any personal information.

Most agencies will accept RESUMES as a supplement to the application. However there are some agencies that will not accept a RESUME. Be sure to find out if your RESUME will be accepted.

If you prepare your RESUME appropriately, it will extend a useful portrait of you to the Oral Board Members.

Your RESUME should be a personal blue print that outlines your, background, experience, education, and qualifications for the position of Firefighter.

Make a photo-copy of your RESUME, it will come in handy as a reference supplement.

RESUMES are useful as a device to use for appraising your accomplishments or examining the direction you are following. Also a RESUME can show you what areas that you need to grow in.

Prior to composing your RESUME, you should begin a file of your accomplishments, etc. Examples:
1. Proof of high school education.
2. College transcripts.
3. Work records.
4. Prior resumes.
5. Prior job applications.
6. Licenses/certificates.
7. Any achievements, experience, education, training, background, credentials, qualifications, prerequisites, etc. that you may have achieved that apply to a Firefighters career.

You do not have to include everything on your RESUME, one of your objectives is to get the Oral Board Members to ask you questions concerning items that you want to discuss.

The RESUME should be designed so that it will get the Oral Board Members attention promptly.

The RESUME should project a mental picture of your background, experience, and qualifications to the Oral Board Members.

You should present all of the material, included on your RESUME, in the past tense.

You should also include your physical condition on your resume.

It is not necessary to include such things as:
1. Marital status. (if you are married, it may be a plus to include this)
2. Height.
3. Weight.
4. Religion.
5. Unimportant information.
6. ETC.

Make sure that you illustrate the fact that you have the ability to comprehend, apply experience, get along with others (esprit de corp), ad-lib, present yourself, work with your hands, be: cautious, dependable, hard worker, durable, enterprising, organized, etc.

Avoid confusion and disorder as this may cause you to overlook useful experience.

Try not to clutter the RESUME, have large margins and open spaces so that the RESUME may be easily read. make the RESUME appealing.

When preparing the RESUME, be totally constructive and candid. Don't include anything that is not true.

It is not necessary to list all of your negative points. Don't highlight your deficiencies.

Try to be direct with all of your information.

Try not to use abbreviations or personal pronouns.

Use verbs to start all of your sentences.

RESUME should be one page, unless the additional information relates directly to the position of Firefighter.

The RESUME should be professionally typed, proof read and spell checked.

RESUMES for the position of Firefighter should be FUNCTIONAL RESUMES as opposed to a CHRONOLOGICAL RESUME.

FUNCTIONAL RESUME: is a resume that will highlight your attributes under four or five function headings. For the position of Firefighter, these headings might be selected from:

1. Objectives.
2. Job skills.
3. Technical skills.
4. Work/related experience/background.
5. Education.
6. Certificates/licenses.
7. References.
8. Personal data.
9. Military.
10. Professional organizations/affiliations.

OBJECTIVES

You may want to submit:

1. Career as a Firefighter in the Fire Service.
2. Entry level Firefighter.
3. Firefighter.
4. A career as a Professional Firefighter that will take advantage of my experience as a Reserve Firefighter and my educational background in Fire Science.
5. Entrance Firefighter, no previous experience.
6. Career in the Fire Service, starting as a Firefighter.
7. ETC.

PERSONAL DATA

Include:

1. Name.
2. Address. (with zip code)
3. Phone numbers. (both work/home, day/night)

JOB SKILLS/TECHNICAL SKILLS/CERTIFICATES

1. EMT-1 certified, or certified Paramedic.
2. Ambulance driving school.
3. Various class driver licenses.
4. Fire systems repair certificate.
5. Building contractors license.

EXPERIENCE

If you have any experience as a Firefighter, such as in the Military, Volunteer Firefighter, Reserve Firefighter, ETC. you should list your experience before you list your education on your resume, otherwise list your education first. In listing your experience, start with the most recent and include positions such as:

1. Volunteer Firefighter.
2. Completion of a Fire Service Training Academy.
3. Fire Cadet with Fire Department.
4. Firefighter in the Military.
5. Ambulance/Paramedic service.
6. Building construction.

EDUCATION

Start with the most recent and include items such as:

1. Degree in Fire Science.
2. Completed such Fire Technology courses as:
   a. Introduction to Fire Technology.
   b. Building and construction.
   c. First Aid.
   d. Emergency Medical Technician I.
   e. Medical terminology.
   f. Fire hydraulics.
   g. Hazardous materials.
   h. Introduction to Arson.
   i. Introduction to Fire Prevention.
   j. Fire tactics and strategy.
   k. Fire protection and suppression.
   l. Fire apparatus and equipment.
   m. Various workshops and seminars.
   o. ETC.
3. County Paramedic Training.
4. Brush fire school.
5. Oil fire school
6. High school graduate.
7. ETC.

MILITARY

Include:

1. Branch of service.
2. Period of active duty.
3. Period of reserve status.
4. Highest rank.
5. Special training:
   a. Firefighting school
   b. Fire drills.
   c. Special teams.

REFERENCES

Have them available and mention that they are available upon request.

LICENSES
(see above)

RESUME/EXAMPLES: (following pages)

RESUME EXAMPLE #1
(Entrance Firefighter, no prior experience)

NAME_____ADDRESS_____PHONE_____

JOB OBJECTIVE: Career in the Fire Service starting as a Firefighter. Strong desire to become a Firefighter using my ability to participate as a team member, along with my self-motivation and conscientious desire to succeed in a career as a Firefighter.

TECHNICAL SKILLS: EMT-1 Certified.
Received training in Fire Protection systems.
Received training in the extinguishment of oil fires.
State Class 2 Drivers License.

EDUCATION: Graduated from _____High School.
Participated in Baseball, Track and Field, Swimming, and was elected Senior Class President.
Currently attending Community College, working towards a degree in Fire Technology.

WORK EXPERIENCE: LIFEGUARD for the City of_____
Completed a training program which stressed physical conditioning, pool and beach safety along with basic First- Aid and rescue techniques. Upon the completion of training was commissioned as a Lifeguard involved in rescue activities and the execution of First-Aid.
(Dates: From_____ To_____)

SERVICE STATION ATTENDANT for a major oil company located in the City. Worked limited hours while attending school. Duties included servicing vehicles with fuel, lubricants and other accessories. Also was permitted to observe and assist mechanics in all types of repair work.
(Dates: From_____To_____)

PERSONAL DATA: Birth date_____, 5'11", 175 LBS.
Married, excellent health, enjoy all sports and mechanical task.
Have gained understanding of Fire Service as a result of visiting various Fire Stations.

RESUME EXAMPLE #2
(related experience and education)
PERSONAL INFORMATION
Name, Address, Phone Number

OBJECTIVE

Desire to secure a position as a Paid Firefighter with the City of_____that will employ experience as volunteer Firefighter and also utilize background of Fire Science education.

WORK EXPERIENCE

VOLUNTEER FIREFIGHTER with the City of_____, from_____ to present. Duties include the maintenance and repair of firefighting apparatus and equipment and maintenance of station. Also participate in department training and respond to emergency incidents and assist in the prevention and control of fires, also assist in saving lives and property. On frequent occasions assume the role of Fire Engineer with the responsibility for the safe and proper operation of departments fire apparatus. On occasion assume the role of Fire Captain and lead other volunteer Firefighters.

AMBULANCE DRIVER with_____Ambulance Service, located within this City, from_____to_____. Responsible for the driving of ambulance while transporting patients in both emergency and non-emergency situations. Also assist in the administering of basic First-Aid and patient management during various emergency and non-emergency incidents. During this period experience was gained in: emergency driving situations, decision making, rescue techniques, city geography, records keeping, responsibility of emergency situations as well as apparatus and equipment.

EDUCATION AND TRAINING

Participated and completed the program for earning Certificate in Firefighting.
Completed Oil Fire school training, date_____.
Attended_____ Community College, from_____to_____.
Earned A.A. Degree in Fire Technology.
Completed E.M.T. 1 training, date_____.
Completed C.P.R. Instructors training, date_____.
Completed advanced First-Aid training, date_____.
Graduated from_____High School, Date_____.

MILITARY

U.S. Navy, from_____ to_____. Honorable discharge.
Completed Military Firefighting School and was a participant of the damage control team. Duties included firefighting, fire drills, etc.

REFERENCES

Excellent references on request.

RESUME EXAMPLE #3
(prior Firefighting experience)

PERSONAL DATA

Name, address, phone # (work/home), age, marital status, height, weight, present health condition.

OBJECTIVE

Employment as a Firefighter with the City of_____.

EXPERIENCE

FIREFIGHTER with the _____ Aircraft Corporation, from _____ to present. Duties include: responding to emergency situations and execute firefighting and/or rescue functions for building, vehicle, or aircraft landing alarms. Also involved in building inspections, dispatching duties, and related training..

AMBULANCE DRIVER with _____ Ambulance Service, from _____ to_____. Duties included responding on emergency and non-emergency incidents and administering basic and advanced First-Aid as needed. Responsible in emergency and non-emergency situations for apparatus and equipment and the care and management of patients.

EDUCATION/TRAINING

Bachelor of Arts Degree in Fire Service Management, from _____ State University, date_____. Associate of Arts Degree in Fire Science, from _____ Community College, date_____. Classes completed include: Fire Protection and Suppression, Fire Apparatus and equipment, Fire Department Management, Introduction to Fire Prevention, Hazardous Materials, Fire Behavior, Tactics and Strategy, Fire Hydraulics, Building Construction, E.M.T.1, Basic and Advanced First-Aid, Medical Terminology, Introduction to The Fire Service, Introduction to U.B.C., Radiological Monitoring, Oral Interview Practice and Procedures, Fire Training Academy, State Certification Classes,
NFPA Firefighter I & II training, Oil Fire School.

CERTIFICATES

Emergency Medical Technician 1
NFPA Firefighter I
NFPA Firefighter II

REFERENCES

References are available on request.

THE COVER LETTER

INTRODUCTION :

Always enclose a covering letter when you mail out a resume. Because your only purpose is to interest an employer in hiring you, your first step must be to get him to read your resume. So keep these facts in mind when writing your cover letter:

1. Address your letter to a specific person by name, when possible.

2. The first twenty words are important; they should attract the reader's interest.

3. Tell your story in terms of the contribution you can make to the employer.

4. Be sure to refer to your resume. It gives the facts.

Prior to constructing your ENTRANCE FIREFIGHTER COVER LETTER AND RESUME you should assemble a supplement to your file, this information should be used in your cover letter. This file should contain examples of:

1. Occasions that exhibit your ability to:
   a. Read and understand written instructions.
   b. To learn.
   c. Make presentations.
   d. Stay calm in stressful situations.
   e. Alertness.
   f. Accept and understand orders.
   g. Be a team member/esprit de corp.
   h. Improvise.
   i. Get along with others.
   j. Utilize knowledge/experience that you have gained.
   k. Mechanical aptitude.

2. Occasions that indicate:
   a. Responsibility.
   b. Motivation.
   c. Integrity.
   d. Sensibility.
   e. Hard work.
   f. Organization.
   g. Dependability.
   h. Durability.
   i. Self confidence.
   j. Physical strength.
   k. Coordination.
   l. Restraint.

COVER LETTER

EXAMPLE #1

Date_____

City of _____
Address_____
City_____ State_____ Zip_____

Dear Sirs,

My name is _____ and I am sincerely interested in becoming a firefighter for the City of _____. I was raised in this City and have progressed through this City's school system. I currently have completed one year at the Community College in this City. I am enrolled in the Fire Technology Program. I am a serious student, which is proven by my "B" average.

In addition to going to College. I work full time in the construction trade as a laborer and a apprentice carpenter. The trade has taught me to be a good hard worker, has improved my physical strength, given me additional self confidence and I have learned much about building construction in general.

I believe that the combination of the education that I have acquired and my work experience makes me a well qualified candidate for the position of Firefighter.

Sincerely,

Signed_____

COVER LETTER

EXAMPLE #2

Date _____

City of_____
Address_____
City_____State_____Zip_____

Dear Sirs,

Let me start with an introduction. My name is_____. I've been employed in the Fire Service for approximately 3 years. My experience varies widely from a maintenance position with the City Fire Department to an Assistant Fire Crew Supervisor for the State Division of Forestry at the Conservation Camp.

I would like to be considered for a position on the County Crew for the upcoming fire season.

I feel my experience with hand crews and pumper crews makes me a highly qualified person to be considered for this position.

You may contact any of my past supervisors to get specific information on my abilities and training in fire suppression.

I will be anxiously awaiting your reply.

                Sincerely yours,

                Signed_____
                Typed name_____
                Address_____
                City, State, Zip_____
                Phone number_____

SUMMARY

As was pointed out in this chapter, there are several important steps to understand when attempting to start a Fire Service career. As you learn about the Firefighters job, it becomes apparent that good study and work habits are important perquisites. It is also apparent that obtaining Fire Service education and training are equally necessary.

When you acquire the necessary education and training there are many organizations that can assist you in finding out when and where the entrance level Firefighter exams will take place.

As we move forward towards our goal we realize that it is extremely important to understand the necessity to properly complete the job application, resume and cover letter. It must be remembered that these components will be the first impression that the oral interview board will have of you. You want to make your first impression a good one.

ASSIGNMENT:

1. Complete application.
2. Develop a one page resume and cover letter.

CHAPTER 3

THE WRITTEN EXAMINATION

PREPARATION

INTRODUCTION:

Usually the first hurdle to pass on your way to becoming a Firefighter is the written examination. In this session we will explore the types of exams that might be encountered. The most prominent type of exams are the general aptitude and the job related examinations.

The general aptitude exams will have questions on such topics as reading comprehension, vocabulary, mathematics, mechanical comprehension, etc. The job related exams will include topics that are part of the Firefighter work environment. Examples of these topics are fire chemistry, fire hose, tools and equipment, ladders, procedures, etc.

When the jurisdiction is using a job related examination, it usually provides a study manual, that the test questions will be taken from. It should be obvious that one way to prepare for the job related examination is to learn as much about the Firefighters job as possible.

Important information to remember on test day is discussed in this session. These helpful hints will help you get through the test with success.

The written portion of the FIREFIGHTERS ENTRANCE LEVEL EXAM may cover several of the various categories of subject matter.

Each jurisdiction will have their own criteria for selection of subject matter. The subject matter will vary from one jurisdiction to the other.

The categories of subject matter that may be included (the complete list may be used or just a portion of the list may be used) in the FIREFIGHTERS ENTRANCE LEVEL WRITTEN EXAM are listed below:
 1. Aptitude.
 2. Reading comprehension.
 3. Verbal ability.
 4. Vocabulary
 5. Spelling.
 6. Grammar.
 7. Mechanical comprehension.
 8. Mechanical ability.
 9. Science.
10. Arithmetic.
11. Math and math concepts.
12. General information.
13. General knowledge.
14. First aid.
15. Emergency Medical Technician techniques.
16. Alphabetical progressions.
17. Number series/progressions.
18. Cubes, figures, matching forms.
19. Pattern analysis.
20. Intelligences.
21. Chemistry.

22. Physics.
23. Judgement.
24. Information included in an entry level Firefighter Candidate Preparation Manual:
    EXAMPLES:
    a. Fire Chemistry.
    b. Fire hose.
    c. Fire Department ladders.
    d. Fire Department tools and equipment.
    e. Ventilation techniques.
    f. Overhaul techniques.
    g. Salvage techniques.
    h. Rescue practices.
    i. Fire Prevention.

HELPFUL EXAM HINTS

1. Arrive early enough to park your car find the location of the place the exam will be given so as to give yourself a chance to relax.
2. Be cooperative with the exam proctor. His/her only function is to help you do your best on the exam.
3. Follow directions.
4. Make note of your exam test number and make sure that you have the correct number of pages in the exam booklet.
5. Read the whole question carefully. Be sure that you know what the question asks and what the choices say.
6. Choose the answer that is generally correct. Answer according to what is generally or usually true, not what would be true in some particular case.
7. Use your time efficiently. The Firefighter written exam is not a speed test, but usually will not give you all the time that you might like to have. In any case, use all the time allowed. If the exam has a time limit, use either the entire time allowed or enough time to go through the entire test twice. Remember the difference between getting a job or not can be as close as one test question.
8. Make decisions. Your decision should be one of the following:
    A. If you know the answer, answer this question now.
    B. If you can figure out the answer, but it will take a lot of time, skip this question and come back to it later. Remember if you skip a question, you must also skip the same number on the answer sheet.
    C. If you do not know the answer and you cannot figure it out, make a guess and answer the question now, unless you have been instructed not to do so. Remember if you are not sure of the answer to a question and you decide to skip it and return to it later, your first intuition is usually the best. Do not change the answer unless you are sure.
9. Don't give up; hang in there and give it and give it a full effort.
10. Try not to change too many of your answers. Remember that the best answer is the one that is usually or generally correct, although if time permits review your exam.
11. Be at your best the day of the exam. Be well rested, allow plenty of time to get to the exam, get there early. Relax the night before the exam.
12. Return your exam and exam materials to the exam proctor.

Some suggestions to follow when taking the FIREFIGHTERS ENTRANCE WRITTEN EXAM:

1. Be cooperative.
2. Follow directions.
3. Make note of your exam test number.
4. Read the whole question carefully; be sure that you know what the question asks and what the choices say.
5. Choose the answer that is generally correct; answer according to what is generally or usually true, not by what would be true in some particular case.
6. Use your time efficiently; The Firefighter written exam is not a speed test, but usually will not give you all the time that you might like to have.
7. Make decisions; your decision should be one of the following:
   A. If you know the answer. (answer this question now)
   B. That you can figure out the answer, but that it will take a lot of time. (skip this question and come back to it later)
   C. That you do not know the answer and that you cannot figure it out. (make a guess and answer this question now, unless you have been instructed not to do so)
8. Don't give up; hang in there and give it a full effort.
9. Try not to change too many of your answers; remember that the best answer is the one that is usually or generally right, although if time permits review your exam.
10. Be at your best the day of the exam; be well rested, allow plenty of time to get to the exam, get there early.
11. Return your exam and exam materials to the exam proctor.

****

Obtain a copy of **"FIREFIGHTER WRITTEN EXAM STUDY GUIDE"** for a complete guide to help you obtain a high score on the written portion of the Firefighters entrance level exam. Available from: INFORMATION GUIDES, P.O. BOX 531, HERMOSA BEACH, CA 90254. 1-800-"FIRE-BKS"

TYPES OF EXAMS/QUESTIONS

The subject matter included in the test is usually presented in the form of a "MULTIPLE-CHOICE EXAM".

MULTIPLE CHOICE EXAMS require that you choose an appropriate answer from a list of answers that have been offered for a question that has been presented. (usually 3 to 5 choices)

The candidate is to choose the one answer that is the best one, the one that most nearly or most often is correct. Sometimes there is no answer that is complete, or exactly correct, or always correct. Choose the best answer. The best answer is the one that is most nearly right, under ordinary conditions.

The candidate must be sure he/she knows what the question asks and what the choices say. On every test, candidates choose wrong answers simply because they failed to read the question carefully, or failed to pay attention to part of the question.

EXAMPLE #1

The questions in a Firefighters multiple choice exam will be:

    A. True-false.
    B. Essay.
    C. Multiple choice.
    D. None of the above.

From the information that you have been given above, you know that "C" is the correct answer from the above choices.

EXAMPLE #2

The number of days in a year is:

    A. 365.
    B. 366
    C. 367
    D. 368.

The answer you should choose is choice A" because it is the one which is most often correct. Choice "B" is true for leap years, but most years have 365 days. Choice "A" is the best answer.

Some of the other types of entry level Firefighter WRITTEN EXAMS are:
1. True-false.
2. Fill in the blanks.
3. Essay test.
4. Matching.

TRUE FALSE QUESTIONS

True and false questions are usually statements that are true or false. You must decide if the statement is accurate (true) or inaccurate (false).

EXAMPLES:

1. When water freezes it expands. (T or F)
   Answer = T.

2. The term used to express electrical pressure is amperage. (T or F)
   Answer = F, correct term is volt.

3. The term which is applied to the distance from the center of a circle to its exterior is radius. (T or F)
   Answer = T.

4. An electric fuse is used to reduce current. (T or F)
   Answer = F, correct use is to protect against electrical overload.

5. The method by which the sun transmits heat to the earth is known as radiation. (T or F)
   Answer = T.

REMEMBER: AVOID QUICK ANSWERS!

FILL IN THE BLANKS QUESTIONS

Fill in the blanks type questions will ask the candidate to fill in a missing word or words so as to correctly complete a sentence or a statement. These type of questions are more difficult to answer, since guessing is eliminated. No answer choice is given, the candidate must recall from memory the correct answer.

EXAMPLES:

1. The freezing point of water on a Fahrenheit thermometer is _____ degrees.
   Answer = 32.

2. A molecule, is the term which is generally applied to the _____ units into which a substance may be divided without destroying it.
   Answer = smallest.

3. Materials which offer a very high or very great resistance to the passage of electricity are known as an _____.
   Answer = insulator.

4. To prevent the passage of gas or air into plumbing fixtures a _____ is used.
   Answer = trap.

5. If substance is soluble in water, it will _____ in water.
   Answer = dissolve.

ESSAY TEST

In an essay test you will be compelled to respond in a written account to a question or statement. (this type of exam is rarely used for entrance level Firefighter exams)

EXAMPLES:

1. Describe SIZE-UP.

    Answer = SIZE-UP is the mental evaluation made by the officer in charge which enables him to determine a course of action. SIZE-UP includes such factors as time, location, nature of occurrence, life hazard, exposure, property involved, nature and extent of fire, available water supply and other firefighting facilities. This report is usually given via radio.

2. Describe WATER, include how it applies to the Fire Service.

    Answer = WATER, (H_2O) is a colorless, odorless, tasteless liquid used widespread for extinguishing fire. Its forms may be liquid, ice (solid), or steam (vapor). The freezing temperature of water is 32 degrees F. The boiling point of water is 212 degrees F. WATER is effective as a cooling agent because it has the largest "latent heat" value of vaporization of all common materials, 90 Btu per pound.

3. Describe CORROSIVE CHEMICALS.

    Answer = CORROSIVE CHEMICALS are chemicals that are particularly injurious to life as, solids, liquids, or gaseous substances they burn, irritate, or destroy organic tissue, most notably the skin and, when taken internally, the lungs and stomach. Include acids, anhydrides, and alkalis.

4. There are three types of radioactive particles, name and rate them in order of danger.

    Answer = The three types of radioactive particles and there danger levels are:
    Alpha, these particles can be stopped by the outer layer of skin.
    Beta, these particles might penetrate to a third of an inch before being absorbed, protective clothing will usually give sufficient protection.
    Gamma, these are the most dangerous, they can go through the body like X-Rays and cause serious injury.

5. What is the method for using a pressure water Fire Extinguisher?

    Answer = The pressure water Fire Extinguisher is most effective when applied close to the fire. Pressure water Fire Extinguishers can be directed from distances of from 30 feet to 40 feet horizontally. It is recommended that the operator should use a continuous stream; keep the nozzle as close to the fire as possible; direct the stream to the base of the fire; use spray or straight stream.

MATCHING QUESTIONS

With matching questions, you will be given two columns of information, each column will have certain details, dates, facts, or a statement. The objective is to accurately merge items from each column together.

EXAMPLES:

1. Match the items in column I with the correct items in column II.

        I   II
- A. Rungs
- B. Screw
- C. Nail
- D. Nut
- E. Hex head bolt

1. Screw driver
2. Hammer
3. Ladder
4. Allen wrench
5. Crescent wrench

Answers = A and 3;   B and 1;   C and 2;   D and 5;   E and 4.

2. Match the correct fire class rating in column 1 with the correct color and shape designation in column 2.

      I   II
- A. Class A fires
- B. Class B fires
- C. Class C fires
- D. Class D fires

1. Red square
2. Yellow star
3. Blue circle
4. Green triangle.

Answers = A and 4;   B and 1;   C and 3;   D and 2.

3. Match the correct fire class rating in column 1 with the correct type of fire in column 2.

      I   II
- A. Class A fires
- B. Class B fires
- C. Class C fires
- D. Class D fires

1. Electrical fires
2. Metal fires
3. Ordinary combustibles
4. Flammable liquids

Answers = A and 3;   B and 4;   C and 1;   D and 2.

4. Match the occupancies in column I with the proper occupancy classification in column II.

I	II
A. Educational.	A occupancy
B. Hazardous	2. B occupancy
C. Assembly	3. E occupancy
D. Business	4. H occupancy
E. Institution	5. I occupancy
F. Carports/fences	6. R occupancy
G. Residences	7. M occupancy

Answers = A and 3; B and 4; C and 1; D and 2; E and I; F and 7; G and 6.

5. Match the proper terms in column I with the corresponding type of tire wear in column II.

I	II
A. Underinflated	1. Tire wear on outside
B. Overinflated	2. Feathering
C. Out of line	3. Flat spots/abrasions
D. Improper balance	4. Flat spots/cupping
E. Heavy braking	5. Wear on one edge
F. Toe in or out	6. Center of tread wear
G. Camber	7. Both edges wear

Answers = A and 7; B and 6; C and 5; D and 4; E and 3; F and 2; G and 1.

INFORMATION SHEET #6

TOPIC: SAMPLE TEST QUESTIONS

INTRODUCTION:
The following examples are typical of types of questions you will encounter on an entrance level Firefighter exam.

EXAMPLE #1

READING:

In any fire, destruction is present in varying degrees. In addition to the destruction caused directly by the fire, there is also destruction caused in fighting the fire. If the sum of destruction to a building by fire and Firefighters is greater for one method of combating the fire than an another, the method causing the lowest level of destruction should be employed.

ACCORDING TO THE ABOVE PARAGRAPH:

 A. The fighting methods are responsible for the major destruction in most cases.
 B. Unavoidable damage by Firefighters should be ignored when choosing a firefighting method.
 C. The aim in the choice of firefighting that causes the least amount of total damage.
 D. Ways of fighting fire should be chosen which are not dangerous to Firefighters.

ANSWER = C.

EXAMPLE #2

SOLVING ARITHMETIC PROBLEMS:

If a gallon of water weighs 8.35 pounds and the weight of a cubic foot of water is 62.5 pounds, then the number of gallons in a cubic foot of water can be found by:

 A. Multiplying 8.35 by 62.5.
 B. Dividing 62.5 by 8.35.
 C. Dividing 8.35 by 62.5.
 D. Subtracting 8.35 from 62.5.
 E. None of the above.

ANSWER = B.

EXAMPLE #3

MAKING COMMON SENSE JUDGMENTS ON PRACTICAL SITUATIONS OR PROBLEMS WHICH MAY ARISE IN FIREFIGHTING OR FIRE PREVENTION:

A Firefighter, while fighting a fire with a hose in a large storeroom, finds it is too dangerous for him/her to remain there any longer. The room is so dark and filled with thick smoke that he cannot see even his/her own hands. He/she can probably best find his/her way out by:

    A. Lighting a match.
    B. Using his memory.
    C. Following the fire hose back from the nozzle.
    D. Spraying water in different directions to locate an exit by sound.

    ANSWER = C.

EXAMPLE #4

UNDERSTANDING HOW MACHINES AND DEVICES WORK:

Of the choices below, which tool would be the best to use for tightening the connections of a household water supply pipe?

    A. Channel locks.
    B. Vice grips.
    C. Crescent wrench.
    D. Pipe wrench.
    E. All of the above.

    ANSWER = D.

EXAMPLE #5

ALPHABETICAL PROGRESSION:

Give the missing letter represented in the blank of the following alphabetical progression:
A C E G ___.

ANSWER = I, Alternate letters of the alphabet have been written in progression.

FIREFIGHTER ENTRY LEVEL CANDIDATE
TEST PREPARATION MANUALS:

There are many jurisdictions that now use these manuals. They will usually hand them out when to you when they accept your application for employment.

These manuals may include any of the previously mentioned subject matter and exam formats.

These manuals are used in exams that will include a reading comprehension section that will assess your capability to hold onto information, recall the information, and then to appropriately interpret this information that has been presented in the manual.

Entrance level Firefighter WRITTEN EXAMS are presented in a way so that they may determine each candidates current level of knowledge along with the candidates capabilities to grasp the functions required for entry level Firefighter.

TEST INFORMATION WILL BE TAKEN DIRECTLY FROM THE INFORMATION PRESENTED IN THE MANUAL!

AFTER THE EXAM

Upon completion of the WRITTEN EXAM make sure that you hand in all of the test matter to the proctor.

After the exam you may ask the proctor as to when you will be notified of the exam results.

After you leave the test location, you should go somewhere that you are comfortable and take the time to list of:
1. The type of questions that were on the exam, ie: multiple choice, true or false, matching, fill-in, etc.

2. The areas that were tested, ie: general knowledge, figures, grammar, mechanical, mathematical, etc.

3. Any question/questions that you feel that you should save on a continual ever-growing list of test questions for your future reference.

4. Anything or feeling that you have about this particular exam that you might want to recall at a later date.

5. Any ideas that you may have on how you could have improved your preparation for this exam.

SUMMARY

The intent of this chapter is to reduce apprehension when taking a Firefighters written examination. Once the student understands the type of examinations that are usually given and how to prepare for them, the test apprehension should be reduced.

Some of the following chapters will revolve around the written examination. Time will be devoted to aptitude testing through various diagnostic exams and you will learn about job related exams through a series of information sheets relating to Fire Department subjects.

ASSIGNMENT:

1. Information Sheet #7: **SITUATION QUESTIONS**, pages 99 - 119.

DIAGNOSTIC TEST

READING COMPREHENSION:

Questions 1 and 2 relate to the following paragraph:

Though the pain from scorpion's sting is intense and may endure for several hours, there is little danger of serious and lasting injury. The scorpion is a sinister-looking creature, with an armored, segmented body supported by eight legs. It possesses numerous eyes, yet for all practical purposes it is devoid of vision, with only monstrous, finger-like pincers to guide it. These pincers are powerful weapons with which the scorpion seizes and crushes its prey. The jointed tail, with its poison needle point, can be wielded with deadly accuracy if the pincers are not effective. As sinister as is the scorpion in the insect world, there is no reason why it should be feared by man. Its enemies are principally insects that destroy grain fields, and in killing the these pest the scorpion serves a beneficent purpose.

1. The title below that best expresses the idea of the above paragraph is:

    A. How the Scorpion Fights.
    B. Enemies of the Scorpion.
    C. A Forbidding Ally.
    D. The Scorpion's Sting.

2. The scorpion's primary threat to its enemies lies in its:

    A. Range of vision.
    B. Speed of motion.
    C. Pointed tail.
    D. Pincers.

3. "Fire drills should be held frequently to be effective. Drills should be so arranged that they will ensure orderly exit under the usual conditions obtained in case of fire. For this reason, drills should be habitually held in unexpected ways and at unexpected times, and should be carefully planned to simulate actual fire conditions." According to this statement, effective fire drills, should be:

    A. Conducted by competent persons only.
    B. Held only under actual fire conditions.
    C. Held at repeated intervals.
    D. Held in unexpected places.

4. "The record of automatic sprinklers in safeguarding life is even more remarkable than the protection of property." This statement illustrates the fact that:

    A. The chief duty of the Firefighter is the protection of life.
    B. The protection of property is not the exclusive function of sprinkler systems.
    C. Property loss due to fire is in inverse proportion to loss of life.
    D. The installation of automatic sprinklers guarantee life rather than property.

Read the next question and then select the answer choice which best summarizes or captures the idea given in the question:

5. Shoring a building means putting up a series of timbers or jacks to strengthen a wall or, if a building has started to collapse, to prevent its further collapse. Shoring does not mean putting walls or floors into their original position. If Firefighters tried to force beams, sections of floors or walls back into place, the building could collapse further and there might be more damage. Temporary shoring is usually all that is done by Firefighters. Building contractors usually do more permanent shoring during the reconstruction of the building.

   A. Firefighters cannot put walls or floors back into their original position, but building contractors are able to do this.
   B. In order to strengthen walls and floors and to prevent further collapse, Firefighters should use timbers or jacks to put walls and floors back into place.
   C. Supporting timbers can be used by Firefighters to strengthen or prevent further collapse of walls or floors of a building.
   D. Building contractors may do both temporary and permanent shoring of buildings.

6. Firehouse "A" has more hoses than Firehouse "B". Firehouse "C" has more ladders than Firehouse "D". Firehouse "D" has more ladders than Firehouse "A" has house. Which statement refers to the largest number of items:

   A. The number of hoses in Firehouse "A".
   B. The number of hoses in Firehouse "B".
   C. The number of ladders in Firehouse "C".
   D. The number of ladders in Firehouse "D".

INSTRUCTIONS FOR QUESTIONS 7 AND 8:

The arrows in the streets tell the direction of one-way streets. All other streets are two-way. Follow the instructions on how to get to the incidents. Assume that you are the Apparatus Driver.

You are responding from the Firehouse located at Agee Street and Hawthorne Avenue. You must obey all one-way signs. The intersections of Graham Street and Hawthorne Avenue is impassable because the Telephone Company is installing new equipment. The Street Department reports that the pavement of Rodgers Avenue between Graham Street and Chanel Street is being torn-up. Therefore, the intersection at Lopez is impassable. The lake is flooding the intersection of Chanel Street and McPherson Avenue

USE THE MAP ON THE FOLLOWING PAGE FOR QUESTIONS 7 AND 8

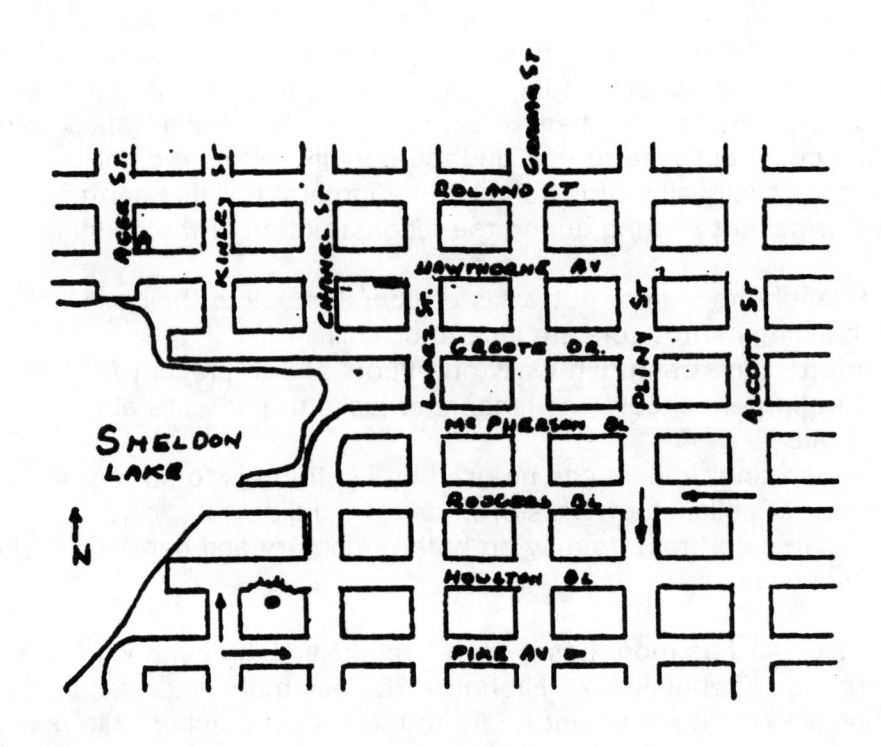

7. There is a fire in the house on Houston Blvd. between Chanel Street and Kinley Street. Which of the following is the shortest route to the fire:

   A. Take Hawthorne To Graham, turn right; take Graham to Houston, turn right; proceed to the fire.
   B. Take Hawthorne to Lopez, turn right; proceed to Houston, turn right; go to the fire.
   C. Take Hawthorne to Chanel, turn right; take Chanel to Houston, turn right; proceed to the fire.
   D. Take Hawthorne to Lopez, turn right; take Lopez to McPherson, turn left; proceed to Graham Street, turn right; proceed to Houston, turn right; go to the fire.

8. After returning to the Firehouse, a person calls to report that a fire hydrant on the corner of Alcott and Groote is flooding the streets. Which of the following is the shortest way to get to the incident.

   A. Take Hawthorne to Lopez, turn left; go to Roland, turn right; take Roland to Alcott, turn right again; go to hydrant.
   B. Take Hawthorne to Alcott, turn right; go to hydrant.
   C. Take Hawthorne to Pliny, turn right; take Pliny to Groote, turn left; go to hydrant.
   D. Take Hawthorne to Lopez, turn right; go to Groote drive and turn left; proceed to hydrant.

82

9. "Much Fire Prevention work can be accomplished through fire company inspections, but it does not reduce the need for specially trained personnel to carry on Fire Prevention and conduct inspections of a technical nature."

   A. Any Firefighter should be able to conduct technical inspections.
   B. Trained technicians should not be used for routine inspections.
   C. Company inspections may reduce the number of technical inspections.
   D. Fire Prevention should be done by company inspections and technical inspections by specially trained personnel.
   E. Technical inspections should be done by specially trained personnel.

10. "United States mortality statistics on the loss of life by fire do not give a true picture because they do not include many types of indirect fire casualties, for example: when forced overboard from a fire at sea. Accidents to Firefighters sustained at fired, and deaths and fatalities following vehicular accidents, and deaths from pneumonia contracted as a result of exposure may not include any mention of fire on the death certificate. All these must be considered as fire casualties." According to the previous paragraph:

    A. Deaths and injuries caused indirectly by fires account for as many as those caused directly by fires, but may not be recorded as fire casualties.
    B. Many deaths reported as drowning, traffic fatalities, pneumonia, and accidents are actually caused indirectly by fire, thus making statistics drawn from these reports inaccurate.
    C. The United States mortality statistics are of little value as far as using these figures in studying deaths from fires is concerned because they do not take into consideration indirect fire casualties.
    D. Information given on death certificates is often inaccurate in that it gives only the immediate causes of death and often omits the real cause.

11. "Employees on duty represent their department to the citizens and are expected to be neat and orderly in their dress at all times." According to this statement, neat and orderly dress of employees while on duty is important because:

    A. Citizens don't care about the appearance of City employees who are off duty.
    B. Employees who are neat and orderly in their dress make better citizens.
    C. If an employee dresses neatly while at work, he will dress neatly when away from work.
    D. People might judge a department by the appearance of its employees.

VOCABULARY

12. REQUISITE means the same as or opposite of:

    A. Stubborn.
    B. Dispensable.
    C. Redeemable.
    D. Pedantic.

13. RECIPROCAL means the same as or opposite of:

    A. Mutual.
    B. Defective.
    C. Residual.
    D. Conditioned.

14. DEFT means the same as or opposite of:

    A. Deaf.
    B. Talkative.
    C. Clumsy.
    D. Valuable.

15. BASHFUL means the same as or the opposite of:

    A. Patient.
    B. Parallel.
    C. Eligible.
    D. Diffident.

16. ADVERSE means the same as or the opposite of:

    A. Admirable.
    B. Opposed.
    C. Detailed.
    D. Rhyming.

17. STEADFAST means the same as or the opposite of:

    A. Irresolute.
    B. Consequential.
    C. Tawdry.
    D. Buoyant.

18. CALLOUS means the same as or opposite of:

    A. Desperate.
    B. Calamitous.
    C. Sensitive.
    D. Hollow.

19. INFAMOUS means the same as or the opposite of:

   A. Dauntless.
   B. Contagious.
   C. Honorable.
   D. Intricate.

20. OBSOLETE means the same as or the opposite of:

   A. Outworn.
   B. Bucolic.
   C. Rampart.
   D. Genuine.

21. DESTITUTE means the same as or the opposite of:

   A. Crazy.
   B. desperate.
   C. Needy.
   D. Injured.

22. ETERNAL means the same as or the opposite of:

   A. Momentous.
   B. Vigilant.
   C. Enraged.
   D. Temporary.

23. PLUMBER is to PIPE as CARPENTER is to:

   A. Blueprint.
   B. Lumber.
   C. Cabinet.
   D. Hammer.

24. PERMIT is to REVOCATION is to MARRIAGE is to:

   A. License.
   B. Divorce.
   C. Death.
   D. Ceremony.

25. PEACH is to FRUIT as COPPER is to:

   A. Penny.
   B. Mining.
   C. Metal.
   D. Silver.

NUMBER ASSOCIATION

26. The next number in the series 2 2 3 3 5 5 8, is:

   A. 8
   B. 10
   C. 9
   D. 11
   E. 12

27. The next number in the series 8 11 9 12 10 13 11, is:

   A. 9
   B. 14
   C. 16
   D. 12
   E. 15

28. The next number in the series 2 4 6 8 10 12, is:

   A. 12
   B. 16
   C. 20
   D. 14
   E. 18

29. The next number in the series 10 8 11 9 12 10, is:

   A. 13
   B. 15
   C. 17
   D. 14
   E. 16

30. The next number in the series 8 24 12 36 18 54 27, is:

   A. 13
   B. 54
   C. 81
   D. 18
   E. 69

31. The next number in the series 22 20 10 12 6 4 2, is:

   A. 2
   B. 6
   C. 10
   D. 4
   E. 8

MATHEMATICS

32. You pay $21 for 3 1/2 tons of coal. What will be the bill for 7 1/2 tons?

    A. $ 9.80
    B. $45.00
    C. $98.00
    D. $42.00
    E. $75.00

33. A farmer buys land for $10,000. He sells it for $12,000, gaining $400 an acre. How many acres did he buy?

    A. 1
    B. 4
    C. 10
    D. 2
    E. 5

34. An artist works on a drawing for 8 weeks and earns $2,000. If he works 5 days a week, how much does he earn every day?

    A. $ 10.00
    B. $ 50.00
    C. $100.00
    D. $ 30.00
    E. $ 80.00

35. If lemons are three for a dime, how much will you pay for a dozen and a half?

    A. 30 cents.
    B. 45 cents.
    C. 60 cents.
    D. 40 cents.
    E. 50 cents.

36. Eight turns of a screw advance it 4 inches. How many inches will ten turns advance it?

    A. 4 1/2
    B. 6
    C. 10
    D. 5
    E. 8

37. A merchant bought chairs at $24.00 a dozen. In selling them he received as much for 2 chairs as he paid for 3 chairs. How much did he charge for 12 chairs?

    A. $25.00
    B. $32.00
    C. $48.00
    D. $30.00
    E. $36.00

38. If you pay $1.20 for two pounds of candy, how much would you pay for half a pound?

    A. 30 cents.
    B. 60 cents.
    C. $2.40.
    D. 45 cents.
    E. 60 cents.

The following three questions, numbered 39, 40, and 41, are to be answered on the basis of the table below. Data for certain categories have been omitted from the table. You are to calculate the missing numbers if needed to answer the questions:

	1956	1957	INCREASE
Firefighters	9326	9744	
Lieutenants	1355	1417	
Captains		443	107
Others			
TOTAL	11469	12099	

39. The number in the "Others" group in 1956 was most nearly:

    A. 450
    B. 500
    C. 475
    D. 525

40. The group which had the largest percentage increase was:

    A. Firefighters
    B. Captains.
    C. Lieutenants.
    D. Others.

41. In 1957, the ratio between Firefighters and all other ranks of the uniformed force was most nearly:

    A. 5 : 1
    B. 2 : 1
    C. 4 : 1
    D. 1 : 1

42. The spring of a spring balance will stretch in proportion to the amount of weight placed on the balance. If a 2-pound weight placed on a certain balance stretches the spring 1/4 inch, then a stretch in the spring of 1 3/4 inches will be caused by a weight of:

    A. 10 pounds.
    B. 14 pounds.
    C. 12 pounds.
    D. 16 pounds.

43. Assume that six men are required to operate a piece of apparatus and each man is on duty 42 hours per week. Assume further that time loss from duty because of vacations, sick leave, and other reasons, amounts to 10 per cent of the companies manpower. Under these conditions, the number of men required to operate this apparatus at full strength around the clock would be most nearly:

    A. 21
    B. 27
    C. 24
    D. 30

44. A machine shop owner sold 13 machine parts for a total of $207.13 (before taxes). What is the price per part:

    A. $12.19
    B. $14.80
    C. $11.93
    D. $15.93

45. Find the taper in 16 feet if the piece of work is tapered 0.18 inches per foot:

    A. 2 inches
    B. 4.13 inches
    C. 2.88 inches
    D. 1.92 inches

46. The end diameters of a piece of work are 6.4 inches and 2.8 inches respectively. Find the taper per inch if the piece is 12.4 inches long (round off to nearest tenth):

    A. .43 per inch.
    B. .29 per inch.
    C. .34 per inch.
    D. .36 per inch.

47. If eleven workmen complete a piece of work in eight days, find the time required by seven men to complete the same piece of work:

    A. 11 days.
    B. 14 1/2 days.
    C. 12.57 days.
    D. 14 days.

48. How many gallons in a 2 1/2 inch hose, 50 feet long:

    A. 16 gallons.
    B. 25 gallons.
    C. 12.75 gallons.
    D. 19.99 gallons.

49. What is the square root of 64:

    A. 6
    B. 8
    C. 7
    D. 9

50. How many cubic inches in a cylinder 1 foot high, 6 inches in diameter:

    A. 900 cubic inches.
    B. 413 cubic inches.
    C. 501 cubic inches.
    D. 339 cubic inches.

MECHANICAL APPTITUDE QUESTIONS:

51. A small gear turns a big gear. Compared to the larger gear, the smaller gear rotates:

    A. At varying speeds.   C. At the same speed.
    B. Faster.   D. Slower.

52. Of the two pulleys below, which one can lift the load more easily:

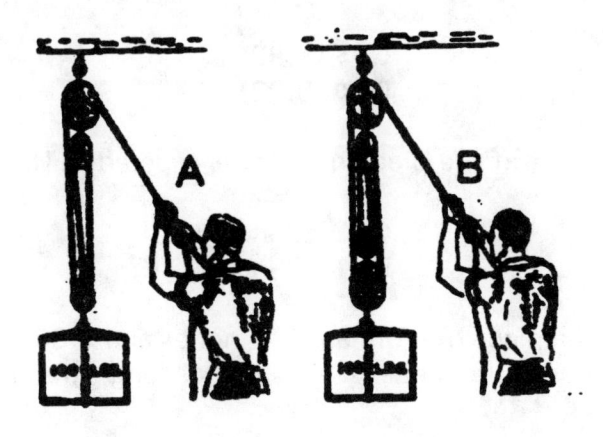

    A. Pulley A.   B. Pulley B.   C. No difference.

Questions 53, 54, and 55 refer to the diagram below which represents a Fire Department hose condition. The numbers 1 and 2 indicate valves. Inlet 3 is connected by means of a hose to one pumper, some distance away, and inlet 4 is connected to another pumper, equally distant from the connection. Outlet 5 leads to the nozzle from which water is thrown on the fire:

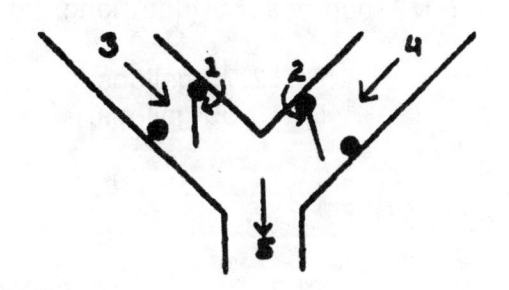

53. The chief purpose of valves 1 and 2 is to:

    A. Reduce the friction between the water and the connection.
    B. Increase the speed of the water in the connection.
    C. Prevent the pumpers from operating at excessive speeds.
    D. Prevent water from one pumper going back into the other pumper.

54. Suppose that both pumpers are operating and water is flowing through the nozzle. Valve 1 is open, but valve 2 is closed. Then, of the following, the most accurate statement is that the pressure at:

    A. Outlet 5 is greater than the pressure at inlet 3.
    B. Inlet 4 is greater than the pressure at outlet 5.
    C. Inlet 3 is greater than the pressure at inlet 4.
    D. Inlet 4 is equal to the pressure at inlet 3.

55. Suppose that the pressure at inlet 4 is much greater than the pressure at inlet 3. On the basis of this information, the most accurate statement is that valve:

    A. 1 will be open and valve 2 will be closed.
    B. 2 will be open and valve 1 will be closed.
    C. 1 and 2 will both be open.
    D. 1 and 2 will both be closed.

56. The following sketch diagrammatically shows a pulley and belt system. If pulley A is made to rotate in a clockwise direction, then pulley C will rotate:

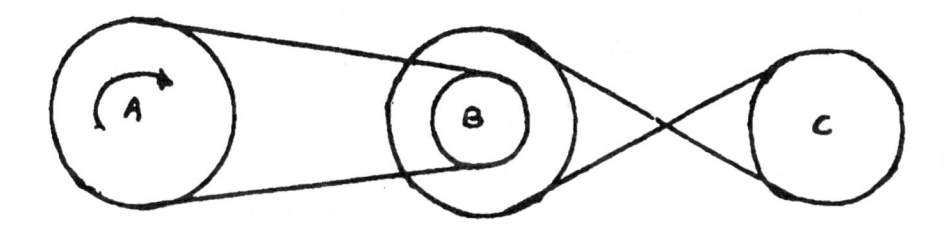

    A. Faster than pulley A and in a clockwise direction.
    B. Slower than pulley A and in a clockwise direction.
    C. Faster than pulley A and in a counter-clockwise direction.
    D. Slower than pulley A and in a counter-clockwise direction.

57. The below sketch shows a weight being lifted by means of a crowbar. The point at which the tendency for the bar to break is greatest at point:

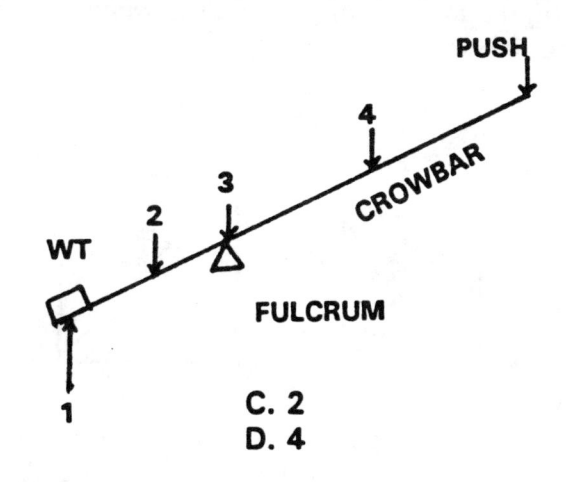

    A. 1
    B. 3
    C. 2
    D. 4

58. The following sketch diagrammatically shows a system of meshing gears with relative diameters as drawn. If gear a is made to rotate in the direction of the arrow, then the gear that will turn fastest is lettered:

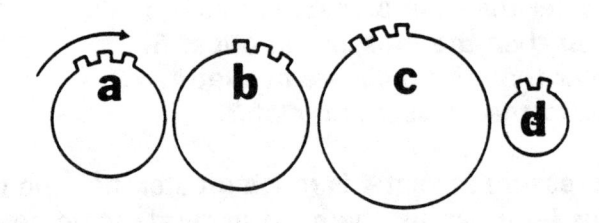

A. Gear "a"   B. Gear "b"   C. Gear "c"   D. Gear "d"

59. An object is to be lifted by means of a system of lines and pulleys. Of the systems shown below, which pulley would require the least force to be used in lifting the weight:

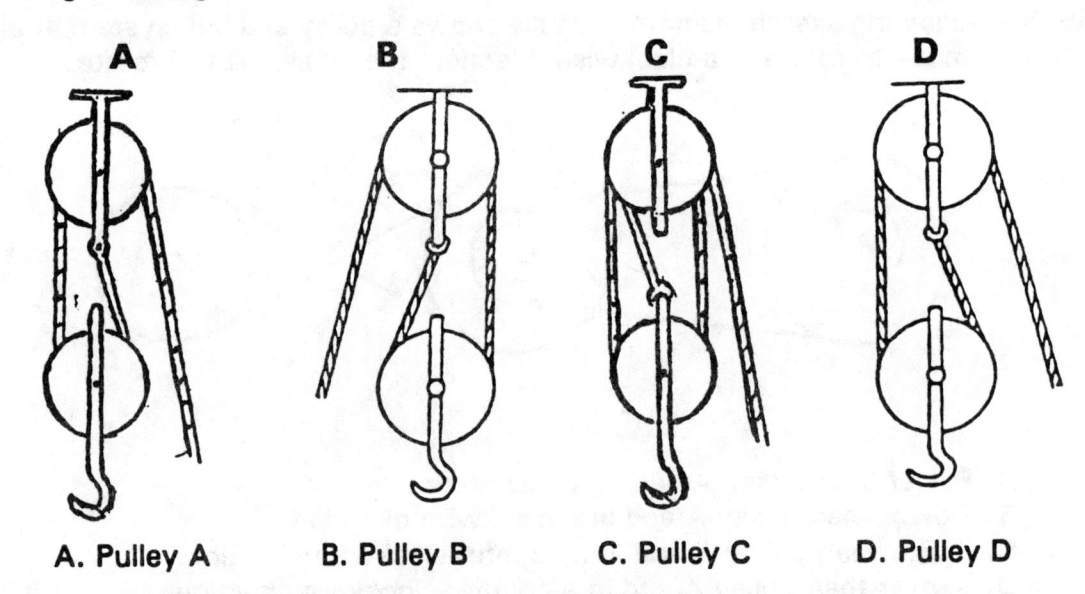

A. Pulley A   B. Pulley B   C. Pulley C   D. Pulley D

60. "When standpipes are required in a structure, sufficient risers must be installed so that no point on the floor is more than 120 feet from a riser."

One of the following diagrams which gives the maximum area which can be covered by one riser is:

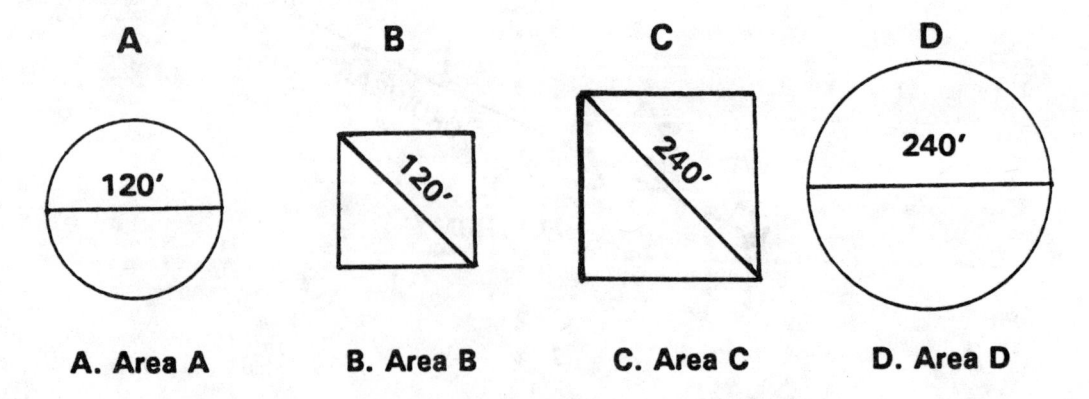

A. Area A   B. Area B   C. Area C   D. Area D

92

CHEMISTRY AND PHYSICS QUESTIONS:

61. The part of the atmosphere that acts as a diluent to the oxygen so as to slow up burning is:

    A. Ozone.
    B. Hydrogen.
    C. Nitrogen.
    D. Dilutogen.

62. The chemical decomposition of a substance by heat is called:

    A. Decompolisis.
    B. Energy output.
    C. Chembustion.
    D. Pyrolysis.

63. The relative density of a vapor or gas as compared to air is called:

    A. Specific gravity.
    B. Vapor density.
    C. Vapor gravity.
    D. Atmospheric pressure.

64. The term used to describe a condition when an oil tank is burning, water is introduced into the tank causing the burning oil to flow out is:

    A. Boil over.
    B. Flow-out.
    C. Oil out.
    D. A screw-up.

65. The lowest temperature of a liquid at which it gives off vapor sufficient to form an ignitible mixture with the air near the surface of the liquid is called:

    A. Fire point.
    B. Flash point.
    C. Ignition point.
    D. Flammable point.

66. An oxidizing agent, a combustive material, and an ignition source are essential for combustion. The previous statement describes the:

    A. Three stages of fire.
    B. Fire tetrahedron.
    C. Fire triangle.
    D. Chemical inhibition.

67. The amount of heat required to raise the temperature of one pound of water one degree Fahrenheit describes:

    A. B.T.U.
    B. Celsius.
    C. Calorie.
    D. Specific heat.

68. One reason for the effectiveness of water, as an extinguishing agent, is that it has a:

   A. Low latent heat rating.
   B. It is plentiful and is low in cost.
   C. High specific heat rating.
   D. All of the above are true.

69. Heat transfer by a circulating medium, either a gas or liquid describes:

   A. Conduction.
   B. Radiation.
   C. Convection.
   D. Biduction.

70. During a fire that gas that is produced from incomplete combustion and is the most dangerous to humans, is:

   A. Hydrogen oxide.
   B. Hyride acid.
   C. Oxygen monoxide.
   D. Carbon monoxide.

DIAGNOSTIC EXAMINATION

EXPLANATORY ANSWERS

1. (D): While the scorpion is sinister looking it kills insects that destroy grain fields.

2. (D): The pincers seize and crush prey.

3. (C): Topic sentence Fire drills should be held frequently to be effective.

4. (B): The statement mentions that sprinklers protect life and property.

5. (C): Topic sentence indicates shoring is used to strengthen and prevent further collapse.

6. (C): Firehouse "C" has more ladders than Firehouse "D", Firehouse "D" has more ladders than Firehouse "A" has hose.

7. (D): The simplest way to solve these problems is to try to get from one point to the other by using the suggested routes in answers: A, B, C, and D.

8. (D): See explanation for #7.

9. (D): Answers A, B, C, and E do not apply.

10. (B): Topic sentence states "statistic on the loss of life by fire do not give a true picture because they do not include many types of indirect fire casualties".

11. (D): Citizens expect employees to be neat and orderly in their dress, therefore the citizens consider the employee as the department.

12. (B): Perquisite means necessary.

13. (A): Reciprocal means mutual.

14. (C): Deft means skilled.

15. (D): Bashful means shy.

16. (B): Adverse means against.

17. (A): Steadfast means constant, firm.

18. (C): Callous means hard.

19. (C): Infamous means bad reputation or wicked.

20. (A): Obsolete means no longer needed.

21. (C): Destitute means needy.

22. (D): Eternal means forever.

23. (B) Carpenter is to lumber. Pipe and lumber are materials to build with.

24. (B): Revocation and divorce both mean to remove or break apart.

25. (C): Peach is part of the fruit family and Copper is part of the metal group.

26. (A): (2 2) (3 3) (5 5) (8 <u>8</u>).

27. (B): 8 11 9 12 10 13 11 <u>14</u>. (8, 9, 10, 11) and (11, 12, 13, <u>14</u>)

28. (C): 2 4 6 8 10 12 <u>14</u>. Add 2 to each number.

29. (A): 10 8 11 9 12 10 <u>13</u>. (10 11 12 <u>13</u>) and (8 9 10)

30. (C): 8x3 24/2 12x3 36/2 11x3 54/2 27x3 <u>81</u>.

31. (D): (22-2 = 20) (10+2 = 12) (6-2 = 4) 2+2 = <u>4</u>.

32. (B): $21/3.5 tons = $6, cost per ton of coal. (7.5 tons x $6) = $45.

33. (E): $12,000 - $10,000 = $2,000 profit; $2,000 profit divided by $400 gained per acre = 5 acres.

34. (B): 5 days per week x 8 weeks = 40 days; $2,000 earned divided by 40 days = $50 earned each day.

35. (C): 3 1/3 cents per lemon; 3 1/3 cents x 18 lemons = almost 60 cents.

36. (D): 8 turns = 4 inches; 4/8 or 1/2 inch per turn; 1/2 x 10 turns = 5 inches.

37. (E) $24 per dozen divided by 12 = $2 per chair; 3 chairs x $2 per chair = $6; 2 chairs = $6; 1 chair = $3; $3 per chair x 12 chairs = $36.

38. (A): $1.20 = 2 pounds; 60 cents = 1 pound; 30 cents = 1/2 pound.

39. (A):

	1956	1957	INCREASE
Firefighters	9326	9744	418
Lieutenants	1355	1417	62
Captains	338	443	107
Others	450	495	45
TOTAL	11469	12099	632

40. (B):
Firefighters: 418/9326 = .0448 or 4.5%
Lieutenants: 62/1355 = .046 or 4.6%
Captains: 107/1355 = .241 or 24.1%
Others: 43/452 = .095 or 9.5%

41. (C): 9744 Firefighters divided by 2355 of all the other ranks = 4 : 1

42. (B): This is a proportion problem. When setting up a proportion problem, the least over the most of one subject (the spring) equals the least over the most of the other subject (the weight):

43. (B): This is another proportion problem. 6 men work 42 hours per week, 24 hours a day for 7 days = 168 hours per week required to operate the apparatus. Ratio = 24 men, 10% Of 24 = 2.4 extra men required; 24 + 2.4 = 26.7 or 27 men are required to operate the apparatus.

44. (D): $207.13 divided by 13 = $15.93

45. (C): 16 feet x .18 = 2.88 inches.

46. (B): 6.4" - 2.8" = 3.6 " total taper.

47. (C): This is a proportion problem. The least over the most of one subject (men), equals the least over the most over the other subject (days) which equals 7 times the unknown = 88; divide 88 by 7 which = 12.57 days.

48. (C): Before this problem can be solved it is necessary to know that:
7.481 gallons = 1 cubic foot and 231 cubic inches = 1 gallon.
The formula to use to find the cubic area of the hose (cylinder), is:
.7854 x (D x D) x the height of the cylinder, therefore:
.7854 x (2.5 x 2.5) x 50 feet =
.7854 x 6.25 x 50' =
.7854 x 6.25 x 600" = 2945 cubic inches.
2945 divided by 231 cubic inches = 12.75 gallons.

49. (A): 8 x 8 = 64; square root of 64 = 8

50. (D): The formula to use to find the cubic area of the cylinder is:
.7854 x (D x D) x height of cylinder.
.7854 x (6 x 6) x 12" = .7854 x 36 x 12 = .339 cubic inches.

51. (B): A smaller gear, because of its size will always make more revolutions than the larger gear.

52. (B): The more pulleys the less force required to move the object.

53. (D): Valves 1 and 2 are basically check valves or one-way valves. Water can only flow one way through the hose connection. Water flow in the opposite direction will force the valves to close.

54. **(C):** Because the water pressure through inlet 3 is greater than inlet 4, which forces valve 2 to close.

55. **(B):** The same principle as #54.

56. **(C):** Pulley C is rotating faster than pulley A because pulley C is rotating at the same speed as pulley B and pulley B is rotating faster than pulley A. Pulley A and B are moving clockwise. Pulley C is moving counter clockwise.

57. **(D):** The fulcrum is where the greatest force is exerted thereby the point at which the bar is most likely to break.

58. **(D):** The smallest gear will always turn the fastest.

59. **(C):** The least amount of pulley use the greater the force required to lift the weight. Pulley A,B, and D would require the greatest force because they are lifting the weight with the least amount of pulley use. Therefore, in figure C, because of more pulley use would require the use of less force.

60. **(C):** If the standpipe is placed in the center of figure C it will be 120 feet from any point to the perimeter.

61. **(C):** The atmosphere is made up mostly of 21% oxygen and 79% nitrogen.

62. **(D):** Pyrolysis is the chemical reaction brought about by heat.

63. **(B):** Definition of vapor density.

64. **(A):** Water is heavier than oil. When it is introduced into a burning oil tank it sinks to the bottom of the tank. As the water heats up and reaches its boiling point of 212 degrees F it boils and forces the burning oil to flow out of the tank.

65. **(B):** Definition of flash point.

66. **(C):** The fire triangle requires fuel (a combustible material), oxygen (oxidizing agent), and heat (ignition source), for combustion.

67. **(A)** Definition of British Thermal Unit (B.T.U).

68. **(C):** C is the best answer because water has a high specific heat rating. It is true that water is plentiful and low in cost, but this fact does not answer the question of its effectiveness. Water does not have a low latent heat rating.

69. **(C):** Definition of convection.

70. **(D):** Carbon monoxide is the most common and hazardous gas produced by incomplete combustion.

INFORMATION SHEET #7

TOPIC: SITUATION QUESTIONS

INTRODUCTION:
Entrance level Firefighter APTITUDE TEST are test in judgement. This category will represent normal Fire Department situations and require candidates to select a course of action that is the accepted norm for the circumstances. EXAMPLES:

SITUATION #1
A friend of yours finds an official Firefighters badge and gives it to his younger brother to use as a toy.

JUDGEMENT OF SITUATION:

SITUATION #2
You and a Firefighter friend of yours are on your way home from a late night movie. You both observe a fire in a hardware store with living quarters on the floors above. Your friend goes to warn the residents. You go to a phone a block away to report fire and then return to the fire scene.

JUDGEMENT OF SITUATION:

SITUATION #3
An on duty Firefighter answers a Fire Department business phone line, but refuses to give the caller his name.

JUDGEMENT OF SITUATION:

SITUATION #4
While a Firefighter is conducting a routine fire inspection on a new business in the city, the owner ask the Firefighter a question concerning a problem that relates to another department within the city that the Firefighter has very little knowledge. The Firefighter suggest to the owner that he contact the appropriate department for his questions.

JUDGEMENT OF SITUATION:

SITUATION #5
During the extinguishment of a fire contained within a U.S. mail box a Firefighter used water to complete the extinguishment of the fire.

JUDGEMENT OF SITUATION:

SITUATION #6
During a fire that was preceded by an explosion, the automatic sprinkler system proved to be ineffective.

JUDGEMENT OF SITUATION:

SITUATION #7
When responding to alarms, many Fire Departments will use pre-established routes.

JUDGEMENT OF SITUATION:

SITUATION #8
Public assembly buildings such as restaurants will have doors that open outwardly.

JUDGEMENT OF SITUATION:

SITUATION #9
Engine company 82 usually conducts company inspections at irregular periods or intervals, but always during normal business hours.

JUDGEMENT OF SITUATION:

SITUATION #10
The company officer refused to turn the gas supply back on, after a house fire, for the home owner.

JUDGEMENT OF SITUATION:

SITUATION #11
While responding to an alarm the fire officer was telling a personal story to the apparatus driver.

JUDGEMENT OF SITUATION:

SITUATION #12
During the training of recruit Firefighters, they will be trained in both engine company and truck company operations.

JUDGEMENT OF SITUATION:

SITUATION #13
During a fire the police department set up "fire lines" so as to keep unauthorized people out of the vicinity while the Fire Department is conducting the appropriate operations.

JUDGEMENT OF SITUATION:

SITUATION #14
While conducting a fire re-inspection a Firefighter observes the owner of the business being inspected correcting a violation during the re-inspection. The Firefighter informs the owner that he will come back at another time to complete the re-inspection.

JUDGEMENT OF SITUATION:

SITUATION #15
At the scene of a fire a rookie Firefighter is given an order from his Captain that is inconsistent with the principles of firefighting that were taught at the fire academy. The Firefighters followed the orders.

JUDGEMENT OF SITUATION:

SITUATION #16
At a house fire the Fire Captain orders is crew to open the doors and windows in a systematic way in order to ventilate the building.

JUDGEMENT OF SITUATION:

SITUATION #17
During a fire inspection a Firefighter suggest to the business owner that he remove large accumulations of combustible trash from the premises.

JUDGEMENT OF SITUATION:

SITUATION #18
During a fire inspection a Firefighter informs the owner of the business that he will have to remove high grass and weeds that are growing near the building.

JUDGEMENT OF SITUATION:

SITUATION #19
During rescue operations involving the use of oxygen, a Firefighter request that family members in the immediate vicinity not smoke.

JUDGEMENT OF SITUATION:

SITUATION #20
While fighting a hot and smoky fire the company officer orders his Firefighters to work in pairs.

JUDGEMENT OF SITUATION:

SITUATION #21
The improper use of oxy-acetylene welders can create a fire hazard because of the failure to control or extinguish hot sparks.

JUDGEMENT OF SITUATION:

SITUATION #22
Fire officers should make sure that their crews are aware that even a single 2 1/2" hose line with a 1" nozzle can deliver more than 2000 lbs of water per minute.

JUDGEMENT OF SITUATION:

SITUATION #23
Fire Department personnel should wear uniforms while on duty.

JUDGEMENT OF SITUATION:

SITUATION #24
A Fire Captain ordered his nozzlemen to attack a fire involved in a group of evenly piled goods from above the fire.

JUDGEMENT OF SITUATION:

SITUATION #25
During company inspections the fire officer instructs his crew to inspect a building that is obviously made of fire resistant construction material such as bricks or natural stone.

JUDGEMENT OF SITUATION:

SITUATION #26
While attempting to get as close to the seat of a very hot fire some Firefighters use a solid object such as a wall to shield themselves from the extreme heat, or they may even cool themselves by wetting down with small streams of water, or by spraying water into the heat waves that are coming from the fire.

JUDGEMENT OF SITUATION:

SITUATION #27
During a pier fire the Fire Captain orders his crew to use sea water as the source of water to fight the fire.

JUDGEMENT OF SITUATION:

SITUATION #28
In attacking a fire on the fifth floor of a building, the Fire Captain orders his crew to advance the hose line up to the fourth floor prior to charging the hoseline with water.

JUDGEMENT OF SITUATION:

SITUATION #29
During a fire, in a warehouse, the commanding officer orders a hole in the roof for ventilation.

JUDGEMENT OF SITUATION:

SITUATION #30
Fire Departments should make it a matter of policy to purchase the best apparatus and equipment and maintain them in the best condition at all times.

JUDGEMENT OF SITUATION:

SITUATION #31
Fire Departments should have the policy that all Firefighters should wear their breathing apparatus in all structure fire situations.

JUDGEMENT OF SITUATION:

SITUATION #32
Prior to responding to an emergency, the Fire Captain orders his apparatus driver to remain in quarters until all of the crew are in their safety gear and sitting down in their seats with seat belts fastened.

JUDGEMENT OF SITUATION:

SITUATION #33
While responding to a pier fire, the Fire captain orders the apparatus driver to avoid driving onto the pier.

JUDGEMENT OF SITUATION:

SITUATION #34
When Fire Departments order new fire apparatus it is a good policy to order apparatus with enclosed cabs.

JUDGEMENT OF SITUATION:

SITUATION #35
While loading hose back on the apparatus, a Fire Captain instructs a rookie Firefighter to avoid bending the hose at places where it has been bent beforehand.

JUDGEMENT OF SITUATION:

SITUATION #36
During a fire in an unoccupied residence a rookie Firefighters finds a large sum of money in a closet and he gives the money to his Captain.

JUDGEMENT OF SITUATION:

SITUATION #37
As an off-duty Firefighter you observe fire apparatus in your rear view mirror, you then pull to the right of the street and stop.

JUDGEMENT OF SITUATION:

SITUATION #38
Upon arrival to a fire scene a Firefighter was ordered by his Captain to take an axe from the apparatus and enter the building at the front door. The Firefighter then knocked down the door for entry.

JUDGEMENT OF THE SITUATION:

SITUATION #39
Most incident commanders will direct their crew to try to contain fires to their point of origin.

JUDGEMENT OF SITUATION:

SITUATION #40
A Fire Captain gives a rookie Firefighter a task to complete. The Firefighter starts the task and then encounters some difficulty in completing the task properly. The Firefighters continues to complete the task improperly.

JUDGEMENT OF SITUATION:

SITUATION #41
A Fire Inspector instructed a school administrator to conduct the schools fire drills on the last Friday of every month just prior to the end of the school day.

JUDGEMENT OF SITUATION:

SITUATION #42
While conducting a company fire inspection, the Fire Captain is told by the owner of the business that the inspection procedure is a waste of time and money. The Fire Captain tells the owner that he is just doing his job and continues the inspection.

JUDGEMENT OF SITUATION:

SITUATION #43
While conducting search and rescue operations in a residence that is on fire, Firefighters usually pay particular attention to closets and spaces under beds and furniture.

JUDGEMENT OF SITUATION:

SITUATION #44
While conducting a company fire inspection the owner of the business ask the Fire Captain a technical question and the Fire Captain does not know the answer to the question. The Fire Captain informs the owner that he is not there to answer questions but to make an inspection.

JUDGEMENT OF SITUATION:

SITUATION #45
As a rookie Firefighter you give an idea for improvement in station maintenance to a fellow Firefighter. At a later date this Firefighter is praised in front of you by your Captain for his excellent idea. You react by telling the Captain the whole story as to where the idea came from.

JUDGEMENT OF SITUATION:

SITUATION #46
During a fire situation the Fire Captain orders his crew to place a hoseline directly in line with the travel of the fire.

JUDGEMENT OF SITUATION:

SITUATION #47
In fire situations, Fire Captains should make it a policy to use the smallest amount of water that is sufficient to put out the fire.

JUDGEMENT OF SITUATION:

SITUATION #48
While fighting a fire in a residence, a Firefighter calls a policeman to remove an individual that keeps trying to re-enter the residence to retrieve some important papers.

JUDGEMENT OF SITUATION:

SITUATION #49
A Firefighters is opening windows in a smoked filled room in such a manner that he starts with the window nearest the entrance and follows the wall around the room until all the windows are open.

JUDGEMENT OF SITUATION:

SITUATION #50
During a fire on the second floor of a building with central air-conditioning the incident commander orders a Firefighter to go to the basement and shut off the air-conditioning.

JUDGEMENT OF SITUATION:

SITUATION #51
As a rookie Firefighter you are told by a fellow Firefighter that the basic reason for fire inspections is so that the Fire Department can exercise its authority on the public.

JUDGEMENT OF THE SITUATION:

SITUATION #52
During a fire in a multi-story building, your Fire Captain orders you to immediately go to the second floor of the building. The order was given prior to any knowledge as to what floor the fire is on.

JUDGEMENT OF SITUATION:

SITUATION #53
A Firefighter is descending a flight of wooden stairs that have been charred by fire. The Firefighter descends backwards keeping close to the wall as he feels each step with his feet.

JUDGEMENT OF SITUATION:

SITUATION #54
In a smoke filled room with no visibility, a Firefighter tries to find a window by standing up and looking into the smoke.

JUDGEMENT OF SITUATION:

SITUATION #55
While fighting a fire in a smoke filled room with no visibility, the hoseline goes limp, no water, the Firefighters leave the hoseline to try and find an exit.

JUDGEMENT OF SITUATION:

SITUATION #56
During a residential fire your Captain orders you to advance a hose line, but not to discharge water until you have actually located the fire.

JUDGEMENT OF SITUATION:

SITUATION #57
After a Fire Captain and his crew has extinguished a house fire he is directed to make an effort to determine the origin of the fire.

JUDGEMENT OF SITUATION:

SITUATION #58
Upon arrival at a fire in a tenement building the Fire Captain of the first in engine company orders his crew to advance a hoseline to the interior stairway of the building.

JUDGEMENT OF SITUATION:

SITUATION #59
While fighting a fire in a warehouse your Captain notices that their are several places of fire origin located within the building, he orders his crew not to move anything after the fire is extinguished.

JUDGEMENT OF SITUATION:

SITUATION #60
When the fire engine arrived on scene, a Firefighter with an axe, jumped off and ran to the door of the building involved and broke in.

JUDGEMENT OF SITUATION

JUDGEMENT OF THE PRECEDING SITUATIONS:

SITUATION #1
A friend of yours finds an official Firefighters badge and gives it to his younger brother to use as a toy.

JUDGEMENT OF SITUATION
The friends action is improper because the badge should be returned to the Fire Department.

SITUATION #2
You and a Firefighter friend of yours are on your way home from a late night movie. You both observe a fire in a hardware store with living quarters on the floors above. Your friend goes to warn the residents. You go to a phone a block away to report fire and then return to the fire scene.

JUDGEMENT OF SITUATION
Proper.

SITUATION #3
An on duty Firefighter answers a Fire Department business phone line, but refuses to give the caller his name.

JUDGEMENT OF SITUATION
The Firefighters action is incorrect because Firefighters should give their name and rank as a matter of routine, when answering a departmental telephone.

SITUATION #4
While a Firefighter is conducting a routine fire inspection on a new business in the city, the owner ask the Firefighter a question concerning a problem that relates to another department within the city that the Firefighter has very little knowledge. The Firefighter suggest to the owner that he contact the appropriate department for his questions.

JUDGEMENT OF SITUATION
The Firefighters suggestion is the proper way to handle this situation.

SITUATION #5
During the extinguishment of a fire contained within a U.S. mail box a Firefighter used water to complete the extinguishment of the fire.

JUDGEMENT OF SITUATION
The use of water to extinguish a fire in this situation is improper because the water may damage the mail that is untouched by the fire, thus making the mail undeliverable.

SITUATION #6
During a fire that was preceded by an explosion, the automatic sprinkler system proved to be ineffective.

JUDGEMENT OF SITUATION
The reason for this is most likely because the explosion may have damaged the pipes that supply the sprinkler system.

SITUATION #7
When responding to alarms, many Fire Departments will use pre-established routes.

JUDGEMENT OF SITUATION
This is a good policy because:
1. Collision of responding apparatus is reduced.
2. Fastest routes are usually pre-planned.
3. Adverse road conditions may be avoided.

SITUATION #8
Public assembly buildings such as restaurants will have doors that open outwardly.

JUDGEMENT OF SITUATION
This is proper because in the event of a fire or other problem, it will prevent people that are exiting from panicking and jamming the doors in a closed position.

SITUATION #9
Engine company 82 usually conducts company inspections at irregular periods or intervals, but always during normal business hours.

JUDGEMENT OF SITUATION
It is proper to conduct fire inspections at various times so that the inspection site will be seen in its normal condition and not in a "ready for inspection condition".

SITUATION #10
The company officer refused to turn the gas supply back on, after a house fire, for the home owner.

JUDGEMENT OF SITUATION
This is the proper procedure to follow because unburned gas may escape from open gas outlets/jets. The local Gas Company will turn the gas back on.

SITUATION #11
While responding to an alarm the fire officer was telling a personal story to the apparatus driver.

JUDGEMENT OF SITUATION
It would be improper to talk to the apparatus driver in this situation, other than to give orders or direction, because it could distract the apparatus driver from concentrating on driving.

SITUATION #12

During the training of recruit Firefighters, they will be trained in both engine company and truck company operations.

JUDGEMENT OF SITUATION

This is a proper procedure because at any fire they may be required to perform the duties of both the engine and ladder company.

SITUATION #13

During a fire the police department set up "fire lines" so as to keep unauthorized people out of the vicinity while the Fire Department is conducting the appropriate operations.

JUDGEMENT OF SITUATION

This procedure is appropriate since it will prevent hindrance with the Fire Departments operations.

SITUATION #14

While conducting a fire re-inspection a Firefighter observes the owner of the business being inspected correcting a violation during the re-inspection. The Firefighter informs the owner that he will come back at another time to complete the re-inspection.

JUDGEMENT OF SITUATION

The Firefighter used good judgement since it would be best to inspect at another time when the violation has been corrected completely.

SITUATION #15

At the scene of a fire a rookie Firefighter is given an order from his Captain that is inconsistent with the principles of firefighting that were taught at the fire academy. The Firefighters followed the orders.

JUDGEMENT OF SITUATION

This was the correct coarse of action for the Firefighter. He should always follow orders in an emergency situation (unless they endanger life) and then at a later time discuss the order that is questionable with his Captain.

SITUATION #16

At a house fire the Fire Captain orders is crew to open the doors and windows in a systematic way in order to ventilate the building.

JUDGEMENT OF SITUATION

This is a proper technique of ventilation so as to:
1. Increase visibility for the Firefighters
2. Reduce the toxic gases.
3. Control the travel of fire and smoke.

SITUATION #17
During a fire inspection a Firefighter suggest to the business owner that he remove large accumulations of combustible trash from the premises.

JUDGEMENT OF SITUATION
This advice is good since any source of ignition could cause the trash to catch fire.

SITUATION #18
During a fire inspection a Firefighter informs the owner of the business that he will have to remove high grass and weeds that are growing near the building.

JUDGEMENT OF SITUATION
This requirement is good because in the event of a fire the high grass and weeds could assist the travel of fire to the building and set it on fire.

SITUATION #19
During rescue operations involving the use of oxygen, a Firefighter request that family members in the location not smoke.

JUDGEMENT OF SITUATION
This is a proper request because the oxygen may cause the cigarette to flare-up dangerously.(oxygen supports combustion)

SITUATION #20
While fighting a hot and smoky fire the company officer orders his Firefighters to work in pairs.

JUDGEMENT OF SITUATION
This is a proper procedure resulting in safer operations since the Firefighters will be able to assist each other in the emergency operations.

SITUATION #21
The improper use of oxy-acetylene welders can create a fire hazard because of the failure to control or extinguish hot sparks.

JUDGEMENT OF SITUATION
This is an accurate statement.

SITUATION #22
Fire officers should make sure that their crews are aware that even a single 2 1/2" hose line with a 1" nozzle can deliver more than 2000 lbs of water per minute.

JUDGEMENT OF SITUATION
This is a true statement and important because of the possibility of building collapse from the weight of the water. (1 LB = 8.35 LBS.)

SITUATION #23
Fire Department personnel should wear uniforms while on duty.

JUDGEMENT OF SITUATION
This is a true statement because other personnel and the public will better be able to recognize personnel that are on duty.

SITUATION #24
A Fire Captain ordered his nozzlemen to attack a fire involved in a group of evenly piled goods from above the fire.

JUDGEMENT OF SITUATION
This is the correct procedure because this is the position of the hose nozzle which will provide maximum water penetration to the goods exposed to the fire.

SITUATION #25
During company inspections the fire officer instructs his crew to inspect a building that is obviously made of fire resistant construction material such as bricks or natural stone.

JUDGEMENT OF SITUATION
This is the correct action to take because the interiors and contents of such a building are susceptible to fire.

SITUATION #26
While attempting to get as close to the seat of a very hot fire some Firefighters use a solid object such as a wall to shield themselves from the extreme heat, or they may even cool themselves by wetting down with small streams of water, or by spraying water into the heat waves that are coming from the fire.

JUDGEMENT OF SITUATION
These are all proper procedures to take so that the Firefighters may get as close to the seat of the fire as possible to allow them to direct their hose streams with accuracy.

SITUATION #27
During a pier fire the Fire Captain orders his crew to use sea water as the source of water to fight the fire.

JUDGEMENT OF SITUATION
This is a proper course of action since this will allow for an unlimited supply of water.

SITUATION #28
In attacking a fire on the fifth floor of a building, the Fire Captain orders his crew to advance the hose line up to the fourth floor prior to charging the hoseline with water.

JUDGEMENT OF SITUATION
This is the proper course of action since the hoseline will be easier to carry and handle prior to charging.

SITUATION #29
During a fire, in a warehouse, the commanding officer orders a hole in the roof for ventilation.

JUDGEMENT OF SITUATION
This a proper course of action since the fire will be fought more effectively by permitting the smoke and hot gases to escape.

SITUATION #30
Fire Departments should make it a matter of policy to purchase the best apparatus and equipment and maintain them in the best condition at all times.

JUDGEMENT OF SITUATION
This is a the best policy since failure of apparatus and/or equipment in emergency situation may have serious consequences.

SITUATION #31
Fire Departments should have the policy that all Firefighters should wear their breathing apparatus in all structure fire situations.

JUDGEMENT OF SITUATION
This would be a good policy since the smoke from all fires are dangerous and may reduce the oxygen content of the air that is being breathed.

SITUATION #32
Prior to responding to an emergency, the Fire Captain orders his apparatus driver to remain in quarters until all of the crew are in their safety gear and sitting down in their seats with seat belts fastened.

JUDGEMENT OF SITUATION
This is a proper procedure to follow because it will reduce the possibility of injury to Firefighters.

SITUATION #33
While responding to a pier fire, the Fire Captain orders the apparatus driver to avoid driving onto the pier.

JUDGEMENT OF SITUATION
This would be the proper procedure to follow since pier fires spread very rapidly and would endanger apparatus and crew.

SITUATION #34
When Fire Departments order new fire apparatus it is a good policy to order apparatus with enclosed cabs.

JUDGEMENT OF SITUATION
This is a good policy because it will protect Firefighters from the possible injury and adverse weather conditions.

SITUATION #35
While loading hose back on the apparatus, a Fire Captain instructs a rookie Firefighters to avoid bending the hose at places where it has been bent beforehand.

JUDGEMENT OF SITUATION
This is a good requirement since repetitive bending of fire hose in the same places will cause weakening of the hose in these locations.

SITUATION #36
During a fire in an unoccupied residence a rookie Firefighters finds a large sum of money in a closet and he gives the money to his Captain.

JUDGEMENT OF SITUATION
This would be the proper procedure for the rookie Firefighter or any Firefighter to follow.

SITUATION #37
As an off-duty Firefighter you observe fire apparatus in your rear view mirror, you then pull to the right of the street and stop.

JUDGEMENT OF SITUATION
This would be the proper course of action to take for off duty Firefighters or any citizen so as to allow the apparatus to pass safely.

SITUATION #38
Upon arrival to a fire scene a Firefighter was ordered by his Captain to take an axe from the apparatus and enter the building at the front door. The Firefighter then knocked down the door for entry.

JUDGEMENT OF THE SITUATION
This was not a proper procedure because the Firefighters should have checked to see if the front door was unlocked.

SITUATION #39
Most incident commanders will direct their crew to try to contain fires to their point of origin.

JUDGEMENT OF SITUATION
This is a good policy because by confining a fire to its area of origin property damage will be minimized.

SITUATION #40
A Fire Captain gives a rookie Firefighter a task to complete. The Firefighter starts the task and then encounters some difficulty in completing the task properly. The Firefighters continues to complete the task improperly.

JUDGEMENT OF SITUATION
This would be an improper procedure, the Firefighter should speak with the Fire Captain about his difficulties prior to completing the task.

SITUATION #41
A Fire Inspector instructed a school administrator to conduct the schools fire drills on the last Friday of every month just prior to the end of the school day.

JUDGEMENT OF SITUATION
This would be a unsatisfactory procedure to follow because fire drills should come at unexpected intervals and times.

SITUATION #42
While conducting a company fire inspection, the Fire Captain is told by the owner of the business that the inspection procedure is a waste of time and money. The Fire Captain tells the owner that he is just doing his job and continues the inspection.

JUDGEMENT OF SITUATION
This would be a poor way to handle this situation, the Fire Captain should explain to the owner the benefits of the inspection program.

SITUATION #43
While conducting search and rescue operations in a residence that is on fire, Firefighters usually pay particular attention to closets and spaces under beds and furniture.

JUDGEMENT OF SITUATION
These are proper procedures since victims, especially children, try to hide from danger in these places.

SITUATION #44
While conducting a company fire inspection the owner of the business ask the Fire Captain a technical question and the Fire Captain does not know the answer to the question. The Fire Captain informs the owner that he is not there to answer questions but to make an inspection.

JUDGEMENT OF SITUATION
This would be a poor response, the Fire Captain should advise the business owner that he does not know the answer but that he will research it and notify him, or tell him where he can get the information.

SITUATION #45
As a rookie Firefighter you give an idea for improvement in station maintenance to a fellow Firefighter. At a later date this Firefighter is praised in front of you by your Captain for his excellent idea. You react by telling the Captain the whole story as to where the idea came from.

JUDGEMENT OF SITUATION
The proper way to handle this would be for you to do nothing about it, but next time make your suggestions to your Captain.

SITUATION #46
During a fire situation the Fire Captain orders his crew to place a hoseline directly in line with the travel of the fire.

JUDGEMENT OF SITUATION
This is a proper procedure since it will increase the chance of controlling the fire.

SITUATION #47
In fire situations, Fire Captains should make it a policy to use the smallest amount of water that is sufficient to put out the fire.

JUDGEMENT OF SITUATION
This is a good procedure to follow, mainly because it will reduce water damage.

SITUATION #48
While fighting a fire in a residence, a Firefighter calls a policeman to remove an individual that keeps trying to re-enter the residence to retrieve some important papers.

JUDGEMENT OF SITUATION
This is the proper course of action to take, because it is the Firefighters responsibility to protect life as well as property.

SITUATION #49
A Firefighters is opening windows in a smoked filled room in such a manner that he starts with the window nearest the entrance and follows the wall around the room until all the windows are open.

JUDGEMENT OF SITUATION
This is a good procedure because it will allow the Firefighter to find his way back to the entrance.

SITUATION #50
During a fire on the second floor of a building with central air-conditioning the incident commander orders a Firefighter to go to the basement and shut off the air-conditioning.

JUDGEMENT OF SITUATION
This would be a proper procedure since this would prevent the spread of smoke by the air conditioning system.

SITUATION #51
As a rookie Firefighter you are told by a fellow Firefighter that the basic reason for fire inspections is so that the Fire Department can exercise its authority on the public.

JUDGEMENT OF THE SITUATION
This is incorrect, the basic purpose of fire inspections is to obtain correction of conditions that create undue fire hazards.

SITUATION #52
During a fire in a multi-story building, your Fire Captain orders you to immediately go to the second floor of the building. The order was given prior to any knowledge as to what floor the fire is on.

JUDGEMENT OF SITUATION
This would be an improper course of action until it is determined it is safe to go to this floor.

SITUATION #53
A Firefighter is descending a flight of wooden stairs that have been charred by fire. The Firefighter descends backwards keeping close to the wall as he feels each step with his feet.

JUDGEMENT OF SITUATION
This is the proper procedure to use because the stairs are presumably better supported next to the wall.

SITUATION #54
In a smoke filled room with no visibility, a Firefighter tries to find a window by standing up and looking into the smoke.

JUDGEMENT OF SITUATION
This would be an unacceptable course of action, the Firefighter should get as low to the floor as he can, since this is where the coolest air will be found, and proceed to a wall and follow the wall until he comes to a window.

SITUATION #55
While fighting a fire in a smoke filled room with no visibility, the hoseline goes limp, no water, the Firefighters leave the hoseline to try and find an exit.

JUDGEMENT OF SITUATION
This would be a poor choice of action, the Firefighters should follow the hoseline back to the outside.

SITUATION #56
During a residential fire your Captain orders you to advance a hose line, but not to discharge water until you have actually located the fire.

JUDGEMENT OF SITUATION
This would be a proper course of action because it will be easier to maneuver the hoseline and it would reduce water damage.

SITUATION #57
After a Fire Captain and his crew has extinguished a house fire he is directed to make an effort to determine the origin of the fire.

JUDGEMENT OF SITUATION
This is a proper procedure because it may help to eliminate this cause of fire in the future.

SITUATION #58
Upon arrival at a fire in a tenement building the Fire Captain of the first in engine company orders his crew to advance a hoseline to the interior stairway of the building.

JUDGEMENT OF SITUATION
This would be a proper course of action because the stairway is a relatively safe and rapid means for evacuating tenants.

SITUATION #59
While fighting a fire in a warehouse your Captain notices that their are several places of fire origin located within the building, he orders his crew not to move anything after the fire is extinguished.

JUDGEMENT OF SITUATION
This would be a good procedure to follow, because several fires starting simultaneously is a strong indication of arson.

SITUATION #60
When the fire engine arrived on scene, a Firefighter with an axe, jumped off and ran to the door of the building involved and broke in.

JUDGEMENT OF SITUATION

The action that the Firefighter took was unwise, the door should have been checked to see if it was open.

INFORMATION SHEET #8

TOPIC: VOCABULARY/VERBAL ABILITY TEST #1

INTRODUCTION:
Entrance Firefighter exams will have VOCABULARY/VERBAL ABILITY TEST to discover a candidates knowledge of words.

When encountering vocabulary test: select the most nearly correct word from among the provided selections. Many times a synonym provided may not be the specific word you would use. But if it is the best of the choices it probably is the correct answer.

Watch out for clues, such as prefixes and suffixes of words. Using the word in a familiar context will help you.

Work rapidly and skim the potential answers to determine promptly the correct choice. Take time out to investigate individual choices only if the words are unfamiliar or troublesome to you.

You will not find exact definitions for the key words. You must use your ability to think and reason in order to select the best answer.

You should know if you are to select a synonym or antonym.

Remember don't panic when you see an unfamiliar word. Most of the words are regularly used in everyday use. Remain calm and you will be able to recognize words and their meanings.

1. **BARGAIN** means most nearly:
A. agreement.
B. debt.
C. routine.
D. design.

2. **ADJOURN** means most nearly:
A. start.
B. handle.
C. attend.
D. complete.

3. **CONSOLE** means most nearly:
A. convey.
B. find.
C. reassure.
D. scold.

4. **MERIT** means most nearly:
A. deficiency.
B. warrant.
C. hope.
D. require.

5. **CORROBORATION** means:
A. accumulate.
B. reduction.
C. confirmation.
D. expenditure.

6. **OPTION** means most nearly:
A. alternative.
B. use.
C. blame.
D. error.

7. **ZEAL** means most nearly:
A. kindness.
B. faith.
C. integrity.
D. enthusiastic.

8. **SLOTH** means most nearly:
A. hatred.
B. lethargic.
C. distress.
D. selfishness.

9. **DEFAMATION** means most nearly:
A. debt.
B. embezzlement.
C. slander.
D. contamination.

10. **PERTINENT** means most nearly:
A. relevant.
B. persuade.
C. foolproof.
D. careful.

11. **IMPROMPTU** means most nearly:
A. laughable.
B. attractive.
C. insulting.
D. spontaneous.

12. **ENDOW** means most nearly:
A. death.
B. abandon.
C. hand down.
D. bless.

13. **ALTERATION** means most nearly:
A. prayer.
B. change.
C. cue.
D. temperance.

14. **ABSORB** means most nearly:
A. consume.
B. spread.
C. purge.
D. repair.

15. **PERCEPTIVE** means most nearly:
A. astute.
B. vicious.
C. dense.
D. fair.

16. **CANDID** means most nearly:
A. correct.
B. direct.
C. hasty.
D. careful.

17. **BANQUET** means most nearly:
A. benefaction.
B. hall.
C. feast.
D. surprise.

18. **INFILTRATE** means most nearly:
A. deflate.
B. rebound.
C. indentation.
D. penetrate.

19. **INTERFERE** means most nearly:
A. stagger.
B. joking.
C. meddling.
D. inquisitive.

20. **THRUST** means most nearly:
A. conclude.
B. pursue.
C. vigil.
D. launch.

21. **FRAIL** means most nearly:
A. fragile.
B. tired.
C. pretty.
D. vain.

22. **HABITUAL** means most nearly:
A. rare.
B. drastic.
C. never.
D. routine.

23. **PROMPT** means most nearly:
A. courteous.
B. punctual.
C. considerate.
D. immaculate.

24. **ADEQUATE** means most nearly:
A. ample.
B. less.
C. too much.
D. too little.

25. **JUSTIFY** means most nearly:
A. shield.
B. appreciate.
C. complete.
D. defend.

26. **INTRUDE** means most nearly:
A. overtake.
b. duplicate.
C. trespass.
D. persist.

27. **INVARIABLY** means most nearly:
    A. sometimes.
    B. periodically.
    C. consistently.
    D. never.

28. **RESOURCES** means most nearly:
    A. portion.
    B. assets.
    C. compromise.
    D. maraud.

29. **WARRANT** means most nearly:
    A. verify.
    B. investigate.
    C. realize.
    D. limit.

30. **OFFENSIVE** means most nearly:
    A. vision.
    B. irate.
    C. unsuitable.
    D. dauntless.

31. **SLENDER** means most nearly:
    A. overgrown.
    B. seductive.
    C. stormy.
    D. thin.

32. **TRANQUILIZING** means:
    A. calming.
    B. supernatural.
    C. solemn.
    D. harmonize.

33. **OBJECTIVE** means most nearly:
    A. competent.
    B. unbiased.
    C. faithful.
    D. compassionate.

34. **CRAFTY** means most nearly:
    A. humorous.
    B. intriguing.
    C. clever.
    D. wicked.

35. **CONTEMPLATE** means:
    A. envision.
    B. appreciate.
    C. invalidate.
    D. disguise.

36. **FALLACY** means most nearly:
    A. clemency.
    B. evidence.
    C. liability.
    D. erroneous.

37. **INTENTION** means most nearly:
    A. desire.
    B. objective.
    C. speculation.
    D. apprehension.

38. **ACUTE** means most nearly:
    A. painful.
    B. sharp.
    C. authentic.
    D. serious.

39. **DEBONAIR** means most nearly:
    A. charming.
    B. dissolute.
    C. dainty.
    D. exorbitant.

40. **HABITAT** means most nearly:
    A. system.
    B. posture.
    C. dwelling.
    D. wardrobe.

41. **GRATIFYING** means most nearly:
    A. offensive.
    B. considerate.
    C. revelation.
    D. pleasing.

42. **CANDID** means most nearly:
    A. swift.
    B. direct.
    C. contention.
    D. kind.

43. **ABBREVIATED** means most nearly:
    A. abridged.
    B. actual.
    C. saturated.
    D. established.

44. **CAMOUFLAGE** means most nearly:
    A. retrieve.
    B. plagiarize.
    C. conceal.
    D. intercept.

45. **INCONSEQUENTIAL** means:
A. unforeseen.
B. trivial.
C. unnecessary.
D. unequivocal.

46. **PROSCRIBE** means most nearly:
A. promote.
B. administer.
C. produce.
D. forbid.

47. **BOUT** means most nearly:
A. boot.
B. match.
C. disease.
D. craft.

48. **CONSOLIDATE** means most nearly:
A. combine.
B. reinforce.
C. identify.
D. classify.

49. **LOQUACIOUS** means most nearly:
A. leafy.
B. redolence.
C. talkative.
D. lightheaded.

50. **EXTOL** means most nearly:
A. praise.
B. release.
C. split.
D. rush.

51. **CONFRONT** means most nearly:
A. challenge.
B. determine.
C. unlock.
D. relinquish.

52. **FORTIFY** means most nearly:
A. excavate.
B. extend.
C. support.
D. reduce.

53. **EXPAND** means most nearly:
A. remodel.
B. increase.
C. construct.
D. absorb.

54. **ACCENTUATE** means most nearly:
A. acquaint.
B. feature.
C. omit.
D. protest.

55. **RELUCTANT** means most nearly:
A. consistent.
B. extreme.
C. disinclined.
D. repose.

56. **LOOMING** means most nearly:
A. threatening.
B. distressed.
C. transient.
D. significant.

57. **BRAVERY** means most nearly:
A. fervor.
B. boldness.
C. independence.
D. potency.

58. **CLIENTELE** means most nearly:
A. attendants.
B. occupants.
C. congregation.
D. patrons.

59. **RELINQUISH** means most nearly:
A. require.
B. surrender.
C. reserve.
D. respite.

60. **ANTECEDE** means most nearly:
A. introduce.
B. reluctance.
C. cavalcade.
D. ceremony.

61. **AMIABLE** means most nearly:
A. offensive.
B. likeable.
C. emotional.
D. confidential.

62. **CORRUPT** means most nearly:
A. majestic.
B. vigil.
C. detached.
D. immoral.

63. **UNCANNY** means most nearly:
   A. remarkable.
   B. juvenile.
   C. dishonest.
   D. spontaneous.

64. **COGNIZANT** means most nearly:
   A. concerned.
   B. unsuitable.
   C. conscious.
   D. thrilled.

65. **PROVOKE** means most nearly:
   A. conceal.
   B. thwart.
   C. vicious.
   D. incite.

66. **ERADICATE** means most nearly:
   A. demonstrate.
   B. remove.
   C. detain.
   D. disguise.

67. **INJURY** means most nearly:
   A. Damage.
   B. threat.
   C. decease.
   D. sorrow.

68. **PROLOGUE** means most nearly:
   A. table of contents.
   B. presentation.
   C. opening.
   D. encasement.

69. **PRECARIOUS** means most nearly:
   A. illusive.
   B. sluggish.
   C. exciting.
   D. risky.

70. **CONTROVERSY** means:
   A. conflict.
   B. dialogue.
   C. presentation.
   D. entertainment.

ANSWER SHEET

VOCABULARY TEST #1

1. A	26. C	51. A
2. D	27. C	52. C
3. C	28. B	53. B
4. B	29. A	54. B
5. C	30. C	55. C
6. A	31. D	56. A
7. D	32. A	57. B
8. B	33. B	58. D
9. C	34. C	59. B
10. A	35. A	60. A
11. D	36. D	61. B
12. C	37. B	62. D
13. B	38. D	63. A
14. A	39. A	64. C
15. A	40. C	65. D
16. B	41. D	66. B
17. C	42. B	67. A
18. D	43. A	68. C
19. C	44. C	69. D
20. D	45. B	70. A
21. A	46. D	
22. D	47. B	
23. B	48. A	
24. A	49. C	
25. D	50. A	

VOCABULARY TEST #2

Another type of vocabulary test will give a word in a sentence and then ask the candidate to select another word from a list that means most nearly the same as the word used in the sentence.

EXAMPLES:

IN THE FOLLOWING SENTENCES THE WORD THAT IS IN **BOLD/CAPITAL** LETTERS MEANS MOST NEARLY? (In these situations only a possible correct choice is given)

1. The breakdown of the machine was due to a defective **GASKET**.

    ANSWER = SEALER

2. The garden contains a **PROFUSION** of flowers.

    ANSWER = ABUNDANCE

3. The person was **CAJOLED** into signing the contract.

    ANSWER = COAXED

4. Noise from the **PNEUMATIC** hammer bothered the man.

    ANSWER = AIR

5. It is impossible to **MISCONSTRUE** my letter.

    ANSWER = MISINTERPRET

6. The mother **ADMONISHED** the child for his behavior.

    ANSWER = WARNED

7. The foreman would not approve the job since it was out of **PLUMB**.

   ANSWER = NOT VERTICAL

8. The speaker's decree was met with general **DERISION**.

   ANSWER = RIDICULE

9. The speaker's decree was **IRRELEVANT**.

   ANSWER = UNCONNECTED

10. The baseball bat was considered a **LETHAL** weapon.

    ANSWER = DEADLY

11. John was appointed **PROVISIONAL** Captain.

    ANSWER = TEMPORARY

12. The Fire Chief used his **PREROGATIVES** in moderation.

    ANSWER = PRIVILEGES

13. The Firefighter used his **INITIATIVE** to complete the task.

    ANSWER = MOTIVATION

14. The Fire Engineer used his **SEASONING** in order to get through the tight situation.

    ANSWER = TRAINING

VOCABULARY TEST #3

And still another type of vocabulary exam will give the candidate a relationship between two words and then ask the candidate to select from a list a relationship from a third given word.

IN THE FOLLOWING SENTENCES, THE WORDS IN **BOLD/CAPITAL** LETTERS ARE RELATED MOST NEARLY TO A WORD IN THE LIST OF GIVEN CHOICES!

1. APRIL is to MONTH as **MONDAY** is to:
   A. minute
   B. hour.
   C. day.
   D. week.

2. ORE is to METAL as **HIDE** is to:
   A. leather.
   B. belt.
   C. plastic.
   D. shoe.

3. GRAIN is to OAT as **VEGETABLE** is to:
   A. wheat.
   B. carrot.
   C. rose.
   D. robin.

4. HUNGER is to NOURISHMENT as **FATIGUE** is to:
   A. work.
   B. play.
   C. sickness.
   D. rest.

5. SCALE is to POUNDS as **RULER** is to:
   A. line.
   B. inches.
   C. length.
   D. measurement.

6. PHYSICIAN is to PATIENT as **LAWYER** is to:
   A. client.
   B. legal representation.
   C. license.
   D. court.

7. FELONY is to CRIME as **EAGLE** is to:
   A. hawk.
   B. fish.
   C. bird
   D. feather.

ANSWER SHEET

VOCABULARY TEST #3

1. C
2. A
3. B
4. D
5. B
6. A
7. C

VOCABULARY TEST #4

Vocabulary exams will also ask the candidate to select the opposite meaning of a word:

1. ALOOF means the opposite of:
   A. mute.
   B. loquacious.
   C. silent.
   D. reserved.

2. PHILANTHROPIC means the opposite of:
   A. grand.
   B. microscopic.
   C. stingy.
   D. powerful.

3. SINISTER means the opposite of:
   A. availing.
   B. unsightly.
   C. factual.
   D. undemanding.

4. AGGRESSIVE means the opposite of:
   A. hostile.
   B. contentious.
   C. tranquil.
   D. forceful.

5. FLAMBOYANT means the opposite of:
   A. crucial.
   B. flashy.
   C. somber.
   D. loud.

6. JADED means the opposite of:
   A. Corrupt.
   B. rejuvenated.
   C. shoddy.
   D. nasty.

7. GROVELING means the opposite of:
   A. cringe.
   B. servile.
   C. admirable.
   D. decline.

8. LOATHE means the opposite of:
   A. despise.
   B. cherish.

ANSWER SHEET
VOCABULARY TEST #4

1. B

2. C

3. A

4. C

5. C

6. B

7. C

8. B

INFORMATION SHEET #9

TOPIC : SPELLING

INTRODUCTION:

Many ENTRANCE LEVEL FIREFIGHTER EXAMS will have a portion of the exam devoted to spelling.

When encountering the spelling portion of an exam: read all the word choices carefully before deciding on your answer. Scan reading is not recommended on spelling test.

Watch out for words that are exceptions to the rules of spelling. Apply as many of the spelling rules that you can and don't let the exception words scare you!

Have a positive attitude and work quickly and methodically. Concentrate on how a word is spelled and not on its definition.

The answers will not follow any special pattern. Each question and answer will be independent of the other questions.

The following examples are characteristic of the degree of spelling that is required on Firefighter exams.

SPELLING TEST #1

CHOOSE THE CORRECT SPELLING OF THE FOLLOWING WORDS:

1.
A. abdoeman.
B. abdomen.
C. abdoman.
D. none of the above.

2.
A. acelerator.
B. accelorator.
C. accelerator.
D. none of the above.

3.
A. accordion.
B. acordyian.
C. accordyon.
D. none of the above.

4.
A. airial.
B. aerial.
C. aereal.
D. none of the above.

5.
A. altornator.
B. altarnator.
C. altanator.
D. none of the above.

6.
A. baffle.
B. bafol.
C. baffol.
D. none of the above.

7.
A. barameter
B. barometor.
C. barometer.
D. none of the above.

8.
A. battalion.
B. batallion.
C. batalloun.
D. none of the above.

9.
A. bimetalic.
B. bimetallic.
C. bymetallic.
D. none of the above

10.
A. bowlyn.
B. bowline.
C. bowlynn.
D. none of the above.

11.
A. campain.
B. campaien.
C. campaign.
D. none of the above.

12.
A. cantilever.
B. cantolever.
C. cantalever.
D. none of the above.

13.
A. captian.
B. captain.
C. captin.
D. none of the above.

14.
A. carpenter.
B. carponder.
C. carponter.
D. none of the above.

15.
A. cardopulmonary.
B. cardyopulmonary.
C. cardiopulmonary.
D. none of the above.

16.
A. dalmation.
B. dalmatian.
C. dalmatain.
D. none of the above.

17.
A. deluge.
B. delouge.
C. daluge.
D. none of the above.

18.
A. difusson.
B. diffusson.
C. diffusion.
D. none of the above.

19.
A. directory.
B. directery.
C. directary.
D. none of the above.

20.
A. durable.
B. dureble.
C. durible.
D. none of the above.

21.
A. elevation.
B. elavation.
C. elivation.
D. none of the above.

22.
A. emergoncy.
B. emergancy.
C. emergincy.
D. none of the above.

23.
   A. endurence.
   B. endurince.
   C. endurance.
   D. none of the above.

24.
   A. expander.
   B. expandor.
   C. expandar.
   D. none of the above.

25.
   A. extinguesh.
   B. extinguish.
   C. extingwish.
   D. none of the above.

26.
   A. flamability.
   B. flammability.
   C. flammibility.
   D. none of the above.

27.
   A. flodlight.
   B. floodlight.
   C. floodlite.
   D. none of the above.

28.
   A. fundamental.
   B. fundemental
   C. fundamentol.
   D. none of the above.

29.
   A. fritcion.
   B. frction.
   C. friction.
   D. none of the above.

30.
   A. fuesible.
   B. fusable.
   C. fusible.
   D. none of the above.

31.
   A. gasoline.
   B. gasoleen.
   C. gasolien.
   D. none of the above.

32.
   A. genorator.
   B. generator.
   C. generrator.
   D. none of the above.

33.
   A. govonor.
   B. governor.
   C. governar.
   D. none of the above.

34.
   A. gravaty.
   B. gravety.
   C. gravity.
   D. none of the above.

35.
   A. gutter.
   B. guttor.
   C. guttar.

36.
   A. hazzard.
   B. hazzord.
   C. hazard.
   D. none of the above.

37.
   A. horizontal.
   B. horazontal.
   C. horizontol.
   D. none of the above.

38.
   A. hydrent.
   B. hydrant.
   C. hydrunt.
   D. none of the above.

39.
   A. hydrolic.
   B. hydralic.
   C. hydraullic.
   D. none of the above.

40.
   A. hydremeter.
   B. hydrameter.
   C. hydrometer.
   D. none of the above.

41.
A. ignightable.
B. igniteable.
C. ignitable.
D. none of the above.

42.
A. ignition.
B. ignetion.
C. ignitian.
D. none of the above.

43.
A. imiscible.
B. immiscible.
C. immiscable.
D. none of the above.

44.
A. immpeller.
B. impellor.
C. impellar.
D. none of the above.

45.
A. impingement.
B. impengement.
C. impingment.
D. none of the above.

46.
A. jaccnife.
B. jackknife.
C. jacknefe.
D. none of the above.

47.
A. janator.
B. janitoor.
C. janitor.
D. none of the above.

48.
A. jeopardize.
B. jepordize.
C. jepardize.
D. none of the above.

49.
A. jostel.
B. jostal
C. jostle.
D. none of the above

50.
A. kerascene.
B. keroscene.
C. kerosene.
D. none of the above.

51.
A. kindaling.
B. kindoling.
C. kindling.
D. none of the above.

52.
A. kinitic.
B. kenitic.
C. kanitic.
D. none of the above.

53.
A. kilogram.
B. kilagram.
C. kilygram.
D. none of the above.

54.
A. knote.
B. knott.
C. knot.
D. none of the above.

55.
A. knowladge
B. knowledge.
C. knowlege.
D. none of the above.

56.
A. labrinth.
B. labyrnth.
C. laberinth.
D. none of the above.

57.
A. latont.
B. latint.
C. latent.
D. none of the above.

58.
A. lattitude.
B. latatude.
C. latitude.
D. none of the above.

59.
  A. legitimate.
  B. legitamate.
  C. lagitimate.
  D. none of the above.

60.
  A. leutenent.
  B. liutenant.
  C. leiutenant.
  D. none of the above.

61.
  A. maxamum.
  B. maximum.
  C. maxumum.
  D. none of the above.

62.
  A. mercontile.
  B. mercantile.
  C. mercuntile.
  D. none of the above.

63.
  A. mezzine.
  B. mezanine.
  C. mezzanine.
  D. none of the above.

64.
  A. millameter.
  B. millemmetor.
  C. millemetre.
  D. none of the above.

65.
  A. multiple.
  B. multaple.
  C. multipel.
  D. none of the above.

ANSWER SHEET

SPELLING TEST #1

1. B	26. B	51. C
2. C	27. B	52. D
3. A	28. A	53. A
4. B	29. C	54. C
5. D	30. C	55. B
6. A	31. A	56. D
7. C	32. B	57. C
8. A	33. B	58. C
9. B	34. C	59. A
10. B	35. A	60. D
11. C	36. C	61. B
12. A	37. A	62. B
13. B	38. B	63. C
14. A	39. D	64. D
15. C	40. C	65. A
16. B	41. C	
17. A	42. A	
18. C	43. B	
19. A	44. D	
20. A	45. A	
21. A	46. B	
22. D	47. C	
23. C	48. A	
24. A	49. C	
25. B	50. C	

SPELLING TEST #2

Another type of SPELLING TEST, will list a group of words, one of which is misspelled. The candidate will be required to select the word from the list that is misspelled.

EXAMPLES:

1. A. aisle   B. cemetary   C. courtesy   D. phlegm

2. A. hypocrisy   B. extraordinary   C. dogma   D. auxilliary

3. A. intorsect   B. launch   C. approach   D. defective

4. A. conferred   B. gigantac   C. synthetic   D. caution

5. A. narrow   B. dispute   C. bargan   D. rapidity

6. A. adjourne   B. hazardous   C. meddling   D. inquisitive

7. A. inquisitive   B. liberate   C. habituall   D. spacious

8. A. reliable   B. boulevard   C. portable   D. illitorate

9. A. permanent   B. abruptly   C. barracade   D. methodical

10. A. punctual   B. penitrate   C. vegetation   D. juvenile

ANSWER SHEET

SPELLING TEST #2

1. ANSWER = B, correct spelling = cemetery

2. ANSWER = D, correct spelling = auxiliary

3. ANSWER = A, correct spelling = intersect

4. ANSWER = B, correct spelling = gigantic

5. ANSWER = C, correct spelling = bargain

6. ANSWER = A, correct spelling = adjourn

7. ANSWER = C, correct spelling = habitual

8. ANSWER = D, correct spelling = illiterate

9. ANSWER = C, correct spelling = barricade

10. ANSWER = B, correct spelling = penetrate

GRAMMAR

INTRODUCTION

Many FIREFIGHTER ENTRANCE EXAMS will have a section which examines your knowledge and understanding of the basic rules of ENGLISH GRAMMAR.

When encountering the grammar portion of an examination: study the instructions and examples carefully. Even though every question will involve your consciousness of basic grammar, the questions will deviate in format.

Always study the sentences carefully. Recognize that the location of a comma can alter the usage.

Watch for sentence fragments or incomplete sentences. They may look correct in a rapid reading, but remember that you are being tested for what is totally correct.

Organize your time and do not spend to much time on any one question. If you can make an intelligent guess, make it or go on to the next question and come back later if you have time.

Remember that you are being requested to consider several things in each sentence, such as, spelling, and punctuation.

Remember to select the best answer. More than one answer may appear to be correct.

The following questions are examples of the components of GRAMMAR that candidates may be tested on, such as punctuation, articulation, and integral parts of speech.

GRAMMAR TEST #1

In the following examples choose the sentence that is the most correct as far as grammar, usage, and punctuation:

1.
   A. Most of the firefighters' training was supervised by captains who's interest lay in training.
   B. Most of the firefighter's training was supervised by captains who's interest lay in training.
   C. Most of the firefighter's training was supervised by captains' whose interest lay in training.

2.
   A. As the reverberations of the siren increase, one of the neighborhood dogs starts to yowl.
   B. As the reverberations, of the siren increase, one of the neighborhood dogs start to yowl.
   C. As the reverberations of the siren increases, one of the neighborhood dogs start to yowl.

3.
   A. Bill yells at Donna that it is her, not he, who hollers.
   B. Bill yells at Donna that it is her, not he, who hollers.
   C. Bill yells at Donna that it is she, not him who hollers.

4.
   A. When a noble fireman trains, he feels real good.
   B. When a noble fireman trains, he feels really good.
   C. When a noble fireman trains, he really feels good.

5.
   A. The candidate asked the proctor what he should do with the examination booklet. Can you envision what he stated?
   B. The candidate asked the proctor what he should do with the examination booklet? Can you envision what he stated?
   C. The candidate asked the proctor what he should do with the examination booklet. Can you envision what he stated.

141

6.
   A. I remember the academy where I trained, while I lived alone, when I was younger.
   B. I remember the academy, where I trained, while I lived alone when I was younger.
   C. I remember the academy, where I trained, while I lived alone when I was younger.

7.
   A. When the staff reports its verdict, some fireman will lose his poise.
   B. When the staff reports their verdict, some firemen will lose there poise.
   C. When the staff reports its verdict, some fireman will lose their poise.

8.
   A. Many of the captains' drills were attended by firefighters who's interest lay in learning.
   B. Many of the captain's drills were attended by firefighters who's interest lay in learning.
   C. Many of the captains' drills were attended by firefighters whose interest lay in learning.
   D. Many of the captain's drills were attended by firefighters whose interest lay in learning.

9.
   A. I asked the instructor what I should do with this examination paper. Can you imagine what he said?
   B. I asked the instructor what I should do with this examination paper? Can you imagine what he said.
   C. I asked the instructor what I should do with this examination paper? Can you imagine what he said?
   D. I asked the instructor what I should do with this examination paper? Can you imagine what he said!

10.
   A. Its in untried emergencies that a firefighter's composure receives its supreme test.
   B. It's in untried emergencies that a firefighter's composure receives its supreme test.
   C. It's in untried emergencies that a firefighters' composure receives its supreme test.
   D. It's in untried emergencies that a firefighter's composure receives its' supreme test.

11.
   A. What you say may be different from me.
   B. What you say may be different from what I say.
   C. What you say may be different than me.
   D. What you say may be different from mine.

12.
   A. It takes study to become a firefighter.
   B. It takes study before you can become a firefighter.
   C. It takes study in becoming a firefighter.
   D. It takes study about to become a firefighter.

13.
   A. You people appreciate we firefighters as much as we firefighters appreciate you.
   B. You people appreciate we firefighters as much as us firefighters appreciate you.
   C. You people appreciate us firefighters as much as us firefighters appreciate you.
   D. You people appreciate us firefighters as much as we firefighters appreciate you.

14.
   A. He use to visit when he was supposed to.
   B. He use to visit when he was suppose to.
   C. He used to visit when he was supposed to.
   D. He visits when he was supposed to.

15.
   A. I saw the engineer and asked him for a hose, fitting, and nozzle.
   B. I saw the engineer, and asked him for a hose, fitting, and nozzle.
   C. I saw the engineer and asked him for a hose, fitting and nozzle.
   D. I saw the engineer asking for a hose, fitting, and nozzle.

16.
   A. A short, old firefighter threw the heavy, soggy, salvage cover.
   B. A short, old firefighter, threw the heavy, soggy salvage cover.
   C. A short old firefighter threw the heavy, soggy salvage cover.
   D. A short old firefighter threw the heavy soggy salvage cover.

17.
- A. We rarely eat meat at our firehouse. My captain being a vegetarian.
- B. We rarely eat meat at our firehouse my captain being a vegetarian.
- C. We rarely eat meat at our firehouse; my captain being a vegetarian.
- D. We rarely eat meat at our firehouse, my captain being a vegetarian.

18.
- A. The commander has only one request. That you respond without delay.
- B. The commander has only one request: that you respond without delay.
- C. The commander has only one request that you respond without delay.
- D. The commander has only one request; that you respond without delay.

19.
- A. The captain insisted that you and he were responsible for the mistakes of Mac and me.
- B. The captain insisted that you and him were responsible for the mistakes of Mac and me.
- C. The captain insisted that you and he were responsible for the mistakes of Mac and I.
- D. The captain insisted that you and him were responsible for the mistakes of Mac and I.

20.
- A. Certainly, even the most experienced firefighter feels stress.
- B. Certainly even the most experienced firefighter feels stress.
- C. Certainly; even the most experience firefighter feels stress.
- D. Certainly: even the most experienced firefighter feels stress.

ANSWER SHEET
GRAMMAR TEST #1

1. C
2. A
3. C
4. C
5. A
6. B
7. A
8. D
9. A
10. B
11. B
12. A
13. D
14. C
15. A
16. C
17. D
18. B
19. A
20. A

GRAMMAR TEST #2

The following examples will show a statement written incorrectly, followed by a space for you to enter the statement correctly.

EXAMPLES:

1. Every country has their representative at the conference.

   _____

2. I read all but the last paragraphs of his essay.

   _____

3. Both the classical and the general course prepares the student for college.

   _____

4. After spending several weeks there, he wrote that he was enamored with the countryside.

   _____

5. The crate of eggs, together with the dozen containers of milk, were punctured in scores of places.

   _____

6. I like those kind of scissors best.

   _____

7. I feel badly because of the oversight.

   _____

8. He was an artist, a scholar, and he could talk!

_____

9. Her brother graduated high school last year.

_____

10. In their own way, everybody has to carry on as best they can.

_____

11. A special light will be required to inspect the engine.

_____

12. The shift before, my captain thinking of other matters thrust his hand into a fire.

_____

13. Not wishing to hurt my captain's feelings, I told him that I was leaving, because I had a previous engagement.

_____

14. Would captain Sutton have survived if he was less imaginative?

_____

15. The worst one of the problems which is confronting me concern morale.

_____

16. When the oral board reports their decision, somebody will earn their promotion.

_____

17. Off in the distance is the fire truck, but there's no signs of it yet.

_____

18. Neither of the paramedics believe that the driver or passenger are alive.

_____

19. Expecting my crew to be on time, their tardiness seemed almost an insult.

_____

20. When mixing it, the foam must be thoroughly blended.

_____

21. It is a thing of excitement, responsibility, and containing satisfaction.

_____

22. If the captain was able, he would demand that the engine return to the station.

_____

23. You first wash the salvage cover in water. Then hang it up to dry.

_____

148

24. The captain sometimes in a good mood gave the study material to others that he had composed.

   _____

25. Even firefighters of uneasy disposition periodically feel an element of primitive enjoyment.

   _____

26. I like those kind of scissors best.

   _____

27. If he would have stated his case, the captain would have been more lenient.

   _____

28. He was a firefighter, writer, and he could talk!

   _____

29. In their own way, everybody has to carry on as best they can.

   _____

30. History has seldom and perhaps will never again record such heroism.

   _____

ANSWER SHEET

GRAMMAR TEST #2

CORRECT STATEMENTS:

1. Every country has its representative at the conference.

2. I read all but the last few paragraphs of his essay.

3. Both the classical and the general course prepare the student for college.

4. After spending several weeks there, he wrote that he was enamored of the countryside.

5. The crate of eggs, together with the dozen containers of milk, was punctured in scores of places.

6. I like this kind of scissors best.

7. I feel bad because of the oversight.

8. He was an artist, a scholar, and a talker.

9. Her brother graduated from high school last year.

10. In his own way, everybody has to carry on as best he can.

11. To inspect the engine, a special light will be required.

12. The shift before, my captain, thinking of other matters, thrust his hand into a fire.

13. Not wishing to hurt my captain's feelings, I told him that I was leaving because I had a previous engagement.

14. Would captain Sutton have survived if he had been less imaginative?

15. The worst one of the problems which are confronting me concerns morale.

16. When the oral board reports their decision, somebody will earn his promotion.

17. Off in the distance is the fire truck, but there are no signs of it yet.

18. Neither of the paramedics believe that the driver or passenger is alive.

19. Expecting my crew to be on time, I regarded their tardiness almost as an insult.

20. When being mixed, the foam must be thoroughly blended.

21. It is a thing of excitement, responsibility, and satisfaction.

22. If the captain were able, he would demand that the engine return to the station.

23. First, wash the salvage cover in water. Then hang it up to dry.

24. Sometimes in a good mood, the captain gave to others the study material that he had composed.

25. Even firefighters of uneasy disposition periodically feel an element of primitive enjoyment.

26. I like this kind of scissors best.

27. If he had stated his case, the captain would have been more lenient.

28. He was a firefighter, writer, and a talker.

29. In his own way, everybody has to carry on as best he can.

30. History has seldom recorded, and perhaps will never again record such heroism.

CHAPTER 4

ENTRANCE FIREFIGHTER ORAL INTERVIEW

ENTRANCE FIREFIGHTER ORAL INTERVIEW

INTRODUCTION

To many, the most difficult part of the Firefighter examination process, is the oral interview. Many people have a difficult time expressing themselves or as they say "tooting their own horn". However that is exactly what must be done in order to obtain a career in the Fire Service. The successful candidate will need to put aside this apprehension and also master the art of organizing his/her thoughts so that the presentation will attract and hold the interest of the oral board. Some ways to improve ones ability to present himself/herself is to practice taking oral interviews, enroll in a public speaking course at the local Community College, and to join a "Toastmaster's Group".

Perhaps the most important key to success in preparing for the oral interview is that it must be understood that preparation begins long before the candidate walks into the room to take the written exam. Some criteria for preparations are to:

1. Learn the specifications and qualifications for the position of Firefighter.
2. Study the principles and philosophy of the profession of Firefighter.
3. Think through each requirement.
4. Correct or improve any weakness you might have.

During the oral interview the candidate must consider his/her appearance and remember that you are a salesperson, the oral board is the client and you are the product. Everything you say will enhance or decrease your chances to become a Firefighter. Therefore it is extremely important to be armed with knowledge of the profession, be positive and alert, think logically, methodically, and with common sense.

This chapter will discuss all the components of the oral interview process from beginning to end, including actual examples of questions.

WHY THEY HAVE ORAL EXAMS

The purpose of an ORAL INTERVIEW is to evaluate each candidates personal qualification, training, experience, attitude, personal attributes, and any intangibles involved that usually outline a candidates likelihood of success or failure for the position of FIREFIGHTER.

The objective of the ORAL INTERVIEW is to determine and identify various elements that have not been examined in the other phases of the exam process.
Some examples are:

1. Adaptability.
2. Attitude.
3. Ability to express yourself.
4. Ability to function under stressful situations.
5. Ability to work with others.
6. Ability to follow directions.
7. Your judgement.
8. Your attitude to working 24 hour shifts.
9. Poise.
10. Stability.
11. Your personal grooming habits.
12. Integrity.
13. Enthusiasm.
14. ETC.

The ORAL INTERVIEW is almost never aimed at testing your professional knowledge. This type of question usually has previously been fulfilled or will be accomplished in the written examination, in the checking of your application/ resume, in reviewing your work record, in reports from your employers, superiors, and associates.

Remember that you asked for this opportunity to be interviewed when you filled for your application. You are not here against your will, and can choose to withdraw at any time.

The decision to be or not to be interviewed is yours to make. You are being interviewed because thus far you have shown the basic qualifications, technical and/or intellectual abilities, along with the experience being sought. You are still in the running and this is an important step in the selection process.

The Oral interview is where the candidate can make up ground if he/she has scored poorly on other portion of the exam process or the candidate can solidify his/her spot on the list if he/she has sailed through the other portions of the exam process.

HAVE A POSITIVE ATTITUDE CONCERNING THE ORAL INTERVIEW, BE ENCOURAGED THAT YOU ARE STILL IN THE RUNNING.

TYPES OF ORAL BOARDS

Entrance level Firefighter oral boards usually consist of people that may be selected from:

1. The community.
2. Business and industry.
3. Fire Department members.
4. Members from other Fire Departments.
5. Protected group representatives.
6. Personnel Department.

Oral interview boards are set up in different areas in different ways and are composed of various groups of interviewers. In some cases the oral boards are composed entirely of members from the Department for which the test is given. Some interview boards are composed of members from other Fire Departments. There is no set standard or criteria as to where, who, how many, or from what positions the board may be chosen.

ORAL INTERVIEWS may be one of two types:

1. Stressful.
2. Non-Stressful.

STRESSFUL ORAL BOARDS will attempt to evaluate you by creating a stressful atmosphere in order to determine how you will perform under this stress. In this interview without introduction you will be asked a question which requires taking a stand on a controversial subject, and then all of the board members will take the opposite view point. The chairman of the board or one of the members previously designated by the board, will inject new controversies on different subjects at random intervals, never letting you fully explain your position on any subject. These types of interviews are seldom used.

NON-STRESSFUL ORAL BOARDS (ordinary interview) are normally of a cordial atmosphere where the board will attempt to put you at ease during the complete interview. This interview usually starts with some type of introductory remark that is designed to reassure you, and is followed by some review questions that you are familiar with. These types of questions are designed to show your interest, vocabulary, along with the volume and tone of your voice, also your mannerisms, etc.

As we have indicated the makeup of the oral board varies from jurisdiction to jurisdiction. Take the time to discover what has been done in the past for Firefighter entrance oral exams in the particular City where the exam is taking place.

HOW TO PREPARE FOR THE ORAL INTERVIEW

You have been preparing for the ORAL INTERVIEW portion of the exam from the time that you started to learn about the Fire Service as a career.

Some of the PRE-ORAL INTERVIEW preparation that you have already established include:

1. Submission of your application and resume. This will have informed the oral board as to your sincerity and desire to obtain the position, along with showing your ability to organize your thoughts, your neatness, and your background. This is the first impression that the oral board will have of you, make it good!
2. Visits to the Fire Stations.
3. Knowledge gained concerning the position.
4. Knowledge gained concerning the department.
5. Knowledge gained concerning the community.
6. Education and training.
7. Knowing your competition.
8. Knowing who is getting hired.
9. Self assessment.
10. Your exam checklist.
11. Practice exams/interviews.

ORAL INTERVIEW EXAM CHECKLIST

1. Know what is on your application.
2. Know what is on your resume.
3. Know the duties of the position.
4. Know the responsibilities of the position.
5. Know the qualifications of the position.
6. Know the Department.
7. Be prepared to present yourself effectively.
8. Know what you have done.
9. Know what you can do.
10. Know what the job requires in the way of performance.
11. Visualize questions that may be asked.
12. Visualize answers to questions.
13. PRACTICE.

EVALUATE MOTIVES FOR PURSUING POSITION

1. Ask yourself why? (the examiners will)

PRACTICE FOR ORAL INTERVIEWS

1. Speak before an audience:
    A. Speech class.
    B. Clubs or organizations.
    C. Use a tape recorder if possible.
    D. In front of a mirror, wife or friends.
2. Orals: Practice orals with qualified officers, firefighters etc.

THE ORAL INTERVIEW

The ORAL INTERVIEW for Entrance Level Firefighter occasionally will last less than 15 minutes, the average time is from 20 to 25 minutes.

On the day of the ORAL INTERVIEW:

1. Arrive early, know where you are going, where you are to park in advance, allow time to get there, park, and reach the interview room, announce your arrival to the proper person, find a place to sit and relax until called.
2. Bring your application for review, or some light reading.
3. Do not bring any exhibits or technical material unless you have been instructed to do so.
4. Be clean and well groomed.
5. Be neat.
6. Avoid all extremes in dress and hair style.
7. Do not wear any pins or emblems.
8. Suit, sport-coat and tie are recommended. For women, a business outfit.
9. While waiting to be called, recheck your attire.
10. While waiting restudy your application.

When you are called for the interview, remember that the interview is a sales interview and that you are the salesman and that the board is the client, and that the product that you are selling is you.

Upon entering the interview room, politely acknowledge any introductions which may be made before you sit down.

If you know a member of the board do not try to hide it, but do not emphasize it either.

Usually the interview board will be seated on one side of a table and you will see an empty chair for you on the opposite side of the table, but there is no set policy regarding the seating arrangements.

Enter the room standing erect and walk with confidence.

You will be introduced to each member of the board. As you are introduced, look each in the eyes and acknowledge along with a firm hand shake if the opportunity is extended. Be confident and direct, use rank titles if applicable.

If the board members are introduced by name, try to remember their names and where possible address the members by name during the interview and/or at its conclusion. Sirs will be satisfactory if you think that you may become confused. Do not confuse title or rank between board members.

You will now be asked to take a seat, sit erect in the chair, do not seem stiff and rigid, be erect and comfortable. Don't shift positions in your seat constantly.

It is alright to gesture with your hands, but not too much! You do not want the board staring at your hands. Do not sit on your hands, put your hands in your lap or on the table in front of you. Be natural in your expression and movements. Avoid distracting movements, such as scratching, pulling on buttons etc.

When asked questions look at the face of the person asking the question as this will help you to focus on the question.

When answering a question, direct the answer to the person that asked the question, but do not ignore the other members of the board.

The board will usually review your application at the start of the interview. Do not interrupt unless there is a significant error made, do not quibble over matters of minor importance.

After the board reviews your application, the first question usually will be a question such as: Tell us something about yourself or Why do you want to be a Firefighter? Try to have some opening statement prepared about yourself, your background and qualification and why you want to be a Firefighter. Remember, never memorize and recite a prepared statement. Have an organized list in your head that you would feel comfortable elaborating on.

Do not give one word answers to questions, try to let the board know why an answer is "yes" or "no", sell yourself!

Answer questions confidently and to the best of your ability. Do not try to "fake it". Be honest, if you do not know an answer to a question tell the board that you do not know the answer. If you lie, you can be disqualified! Be sincere and deliberate with your answers.

Don't worry if you are nervous, the board expects this. Board members want you to do your best. The board members are not your enemies, they are just observers looking for the best candidate for the position. Let the board know that you are the best candidate for the position.

Remember that you want this position, sell yourself to the board, don't sell yourself short. The board members are trying to select the best person for the position, give them something to work with.

Let the board know that you are a serious candidate. Act professional at all times during the interview. If appropriate a little humor is alright, but do not overdo it. Be attentive.

When answering questions:

1. Modulate your voice, don't speak in a monotone manner.
2. Speak-up, but not too loudly.
3. Make sure that you understand the question before answering, if not, restate the question or ask for a clarification. Don't overdue this, do not make the board repeat every question.
4. Be honest.
5. Be pleasant, smile occasionally.
6. Wait until the entire question is asked, don't interrupt the questions.
7. Do not argue. If you make a stand on an issue, do not change your opinion unless you are proven wrong.
8. Answer questions completely and then stop. Don't try to dominate the interview.
9. Expect abrupt changes in the questioning.
10. Reply with your answers promptly, not hastily.
11. Don't spend too much time praising your current job, you are looking for a new career.

Sometimes a board member may stop you in the middle of answering a question, this could be a positive indication that he feels that you have already answered the question to his liking and he wants to cover another area. Never try to continue to answer the question! Be ready for the next question.

Just prior to the end of the interview, the board will usually ask if there is anything you would like to add:

1. If they have overlooked one or more of your strong points, you should bring it out now.
2. Do not start an extended presentation now.
3. If you have nothing to add, just say; No thank you, I believe we have covered everything.
4. Don't compliment the board members.

The board will let you know when the interview is over, at this time just thank them, shake their hands and leave the room. Do not make a speech at this time. The only time that you should say anything is if something very important has been left out, then you should briefly mention it. You actually terminate the interview, at this point you can actually talk yourself out of a good impression or fail to present an important bit of information.

When you leave the interview room you should leave with poise and confidence watch for an offer of a handshake, if an offer is made, shake hands.

HOW TO PUT YOUR BEST FOOT FORWARD

Throughout this process, you may feel that the oral board is trying to penetrate your defenses, to seek out your hidden weaknesses, and to try to embarrass or confuse you. This is not the case, they are compelled to make an evaluation of your qualifications for the position. They want to see you at your best.

The oral board must interview each candidate and a noncooperative candidate may become a failure in spite of the boards best efforts to bring out the candidates qualifications. You must put your best foot forward, the following are some suggestions on how to accomplish this task:

BE NATURAL: keep your attitude confident, but not cocky. If you are not confident that you can do the job, don't expect the board members to be. Don't apologize for your weaknesses, bring your strong points. The board is interested in a positive, not a negative presentation. Cockiness will antagonize any board member, and make him wonder if you are covering up a weakness by a false show of strength.

GET COMFORTABLE: don't lounge or sprawl, sit erectly but not stiffly. A careless posture may lead the board to conclude you are careless in other things, or at least that you are not impressed by the importance of the interview. Don't fuss with your clothing or with a pencil or ash tray, etc. Your hands may occasionally be useful to emphasize a point; don't let them detract from your presentation by becoming a point of distraction.

DON'T WISECRACK: or make small talk, this is a serious situation, and your attitude should show that you consider it as such. Further, the time of the board is limited; they don't want to waste it, and neither should you.

DON'T EXAGGERATE: your experience or abilities, the board has your application in front of them and also from other sources may know more about you than you think. Also you probably won't get away with it anyway. An experienced board is rather adept at spotting such a situation. Don't take a chance!

IF YOU KNOW A MEMBER OF THE BOARD: don't make a point of it, but don't hide it. Certainly you are not fooling him, or the other members of the board. Don't try to take advantage of your acquaintance with this individual, it will likely hinder your score.

DON'T ATTEMPT TO DOMINATE THE INTERVIEW: Let the board have control. They will give you clues, don't assume that you have to do all of the talking. Be aware that the board has a number of questions to ask you. Don't try to take up all the interview time by showing off your extensive knowledge of the answer to the first question that you are asked.

BE ATTENTIVE: you will have a limited time in the interview, keep your attention at a sharp level throughout this period. When a board member is questioning you, give this person your undivided attention. Direct your response primarily to the board member that ask the question, but don't exclude the other board members.

DON'T INTERRUPT: A board member may be stating a problem for you to analyze. When the time comes you will be asked a question. Let the board member state the problem, and wait for the question.

MAKE SURE THAT YOU UNDERSTAND THE QUESTION: Don't try to answer until you are sure what the question is. If the question is not clear, restate it in your own words or ask the board member to clarify it for you. Don't haggle about minor elements.

REPLY PROMPTLY: But don't reply hastily. A common entry on oral board rating sheets is "candidate responded readily" or "candidate hesitated in replies". Respond promptly and quickly, but don't jump to hasty, ill-considered responses.

DON'T BE PREEMPTORY IN YOUR ANSWERS: A brief answer is proper, but don't fire your answers back. This is a losing game from your point of view. The board member can probably ask questions much faster than you can answer them.

DON'T TRY TO CREATE ANSWERS THAT YOU ASSUME THE BOARD WANTS TO HEAR: The board is interested in what kind of a mind you have and how it works, not in playing games. Most board members can spot this tactic and will grade you down for it.

DON'T CHANGE VIEWS IN ORDER TO PLEASE THE BOARD: Board members will take a contrary position in order to draw you out and see if you are willing and able to defend your point of view. Don't start a debate, but don't surrender a good position. If your position is worth taking, defend it!

IF YOU ARE SHOWN TO HAVE MADE AN ERROR IN JUDGEMENT, DON'T BE AFRAID TO ADMIT THE ERROR: The board knows that you are forced to reply without any opportunity for careful consideration. Your answer may be demonstrably wrong. If so, admit it and get on with the interview.

DON'T SPEND TOO MUCH TIME DISCUSSING YOUR PRESENT JOB: The opening statements may concern your present employment. Answer the question but don't go into an extended discussion. You are being examined for a new job, not your present one. As a mater of fact, try to phrase all you answers in terms relating to the position of Firefighter.

DON'T TELL STORIES: Keep your responses to the point. If you feel the need for illustration from your personal experience, keep it short. Leave out the minor details. Make sure that the incident is true.

DON'T BE TECHNICAL OR BORING: The board is not interested in ponderous technical data at this time.

DON'T USE SLANG TERMS: many a good response has been weakened by the injection of slang terms or other jargon. Oral boards usually will notice any slips of the grammar or any other evidence of carelessness in speech habits.

DON'T BRING DISPLAYS OR DEMONSTRATIONS: The board members are not interested in letters of reference etc.

WAYS TO STRIKE OUT IN AN ORAL INTERVIEW

1. Poor personal appearance.
2. Lack of interest and enthusiasm, appear lazy.
3. Passiveness or indifference.
4. Overemphasis on wages.
5. Condemnation of past employers.
6. Failure to look at board members during interview.
7. Limp, fishy handshake.
8. Indefinite response to questions.
9. Overbearing, overaggressive, conceited with superiority or "know it all" attitude.
10. Inability to express self clearly: poor voice, diction, and/or grammar.
11. Lack of planning for Firefighters position.
12. Lack of confidence and poise: nervous, ill at ease.
13. Make excuses: evasive; hedges on unfavorable factors in work record, etc.
14. Lack of tact, courtesy; ill mannered.
15. Lack of maturity and/or vitality.
16. Indecision.
17. Sloppy application.
18. Merely "shopping" for the position.
19. Want position for only a short time.
20. Lack of interest in jurisdiction.
21. Domination of interview, high pressure type.
22. Low moral standards.
23. Intolerant, strong prejudices.
24. Narrow interest.
25. Inability to listen, and/or take criticism.

AFTER THE ORAL INTERVIEW

After you are excused form the ORAL INTERVIEW, go somewhere that you feel comfortable and make a list of:

1. How you felt during the interview.
2. Any question/questions that you feel you may want to recall at a later date.
3. Types of questions asked.
4. How questions were asked.
5. How the oral board reacted to your answers.
6. How you may have improved your preparation for this exam.

ORAL INTERVIEW QUESTIONS

1. Tell us about yourself.

2. In order that we may all feel more comfortable, even though we have your application/resume in front of us, and have reviewed it, please give us a brief history of your background.

3. How have you prepared yourself for the position of Firefighter?

4. Why do you want the position of Firefighter?

5. What are your qualifications for the position of Firefighter?

6. How do you feel that hiring you will improve this Fire Department?

7. Why should you be one of the candidates hired to the position of Firefighter on this Fire Department?

8. What do you consider to be your greatest strength?

9. How would you describe yourself?

10. How would your peers describe you?

11. Why did you choose the Fire Service for your career?

12. In what kind of work environment are you the most comfortable?

13. How do you work under pressure? Can you give some examples of when you had to work under pressure?

14. Describe the most significant contribution you have made in one of your past jobs, and how has this prepared you to be a Firefighter?

15. What motivates you to put forth your greatest effort?

16. Do you honestly feel that you have the credentials to succeed as a Firefighter? Why?

17. What do you think it takes to be successful in the Fire Service?

18. How do you determine or evaluate success?

19. What unique characteristic do you possess that would enhance you in the role, as Firefighter, as compared to your competitors?

20. What makes you think that you are more capable of handling this job than any of your competitors?

21. What do you think your greatest problem as a Firefighter for this Fire Department will be?

22. What major problems, in the Fire Service, do you expect to encounter, and how do you think you will deal with it?

23. The position of Firefighter demands that you accept a great deal of responsibility. In this respect, what is your primary weakness? How are you overcoming this weakness?

24. What is the technique that you use for solving problems?

25. Have you considered the hazardous nature of the work performed as a member of the Fire Department?

26. Describe what you believe makes up a Firefighters daily routine:

27. What would you do if you knew that some of the members in your company disliked you?

28. What two or three of your accomplishments have given you the most satisfaction? Why?

29. What would be your first goal, after your are selected for the position of Firefighter?

30. What are your long range goals in the Fire Service? How do you plan to achieve these goals?

31. How would your peers describe you in your present job?

32. What do you see yourself doing in the Fire Service five (5) years from now?

33. Describe your most rewarding job experience with your present or past employment.

34. Of all the courses that you have taken in College and/or special seminars, which do you feel has been the most beneficial to you?

35. What part of your formal education do you think will help you the most as a Firefighter?

36. What type of community activities have or are you involved in?

37. What is your opinion of **WOMEN** in the Fire Service?

38. Are you aware that there is a probationary period for "rookie" Firefighters?

39. Why have you chosen the Fire department as a career rather than the Police department?

40. You are a member of the department and a fellow Firefighter has been injured fighting the fire. There are citizens who are also injured at the fire. To whom would you give your first attention relative to first aid?

41. You are a member of the Fire Department and you suspect that a fellow member is pulling false alarms. What would you do about it?

42. As a member of the Fire department, what action would you take when your superior officer asked you relative to the drinking habits of a fellow member of the department?

43. If you are in the Fire Station and time is hanging on your hands, what would you do to occupy your standby time?

44. You are in a hotel lobby and fire breaks out. Would you notify the hotel clerk to send in an alarm, or would you run out and pull the nearest fire alarm box.

45. Is there any area that has not been touched on, or is there anything that you would like to add?

ADDITIONAL SITUATION QUESTIONS

1. What would you do if you were responding to a fire incident and your Captain told you to turn right and you knew the direction was straight ahead?

2. What would you do if you were driving to an emergency and your Captain told you to drive faster than you think it is safe to drive?

3. While you are maneuvering your apparatus at an emergency scene, you accidentally hit a parked car. **NO ONE SAW YOU** hit the car! what will you do?

4. If you felt that your Captain is issuing the wrong orders, how would you confront him?

5. How would you handle a situation in which a Firefighter is consistently misusing equipment?

6. After a fire, you are wiping down some equipment while you are still on the scene. A thankful property owner offers you a beautiful antique model fire engine, worth several hundred dollars. At first you refuse but the person insists, saying that he was just going to dispose of it anyway. He then places it on the fire apparatus. What would you do?

7. As a Firefighter you are eating dinner at a restaurant and notice that the owner has locked the required exits. What would you do?

8. As a Firefighter you are performing overhaul and observe another Firefighter putting something into his turnout pocket, what would you do?

9. The Firefighter that you relieve on a regular basis continually leaves the equipment in less than the minimum operational condition. What actions, if any would you take to correct this recurring problem?

10. You are told to respond as a member of a task force to a major brush fire about 100 miles away. You are ordered to respond on an apparatus that you think is obviously unsafe and unable to make the trip. How would you handle this situation?

11. If your Captain asked you to move a fire apparatus to another location, at an emergency scene, and while you are maneuvering the apparatus you accidentally hit a parked car. No one saw you hit the car, what action will you take, if any?

12. You are riding down the street and notice that a gasoline tanker truck is leaking quite a bit of gasoline, what would you do in this situation?

13. Assume that you are a newly appointed Firefighter and you make a mistake in your work, that if discovered might cause embarrassment and annoyance to your supervisors, how would you handle this situation?

14. Assume that you are a younger less experienced Firefighter at a station with another Firefighter that constantly tells you that your ideas for improving efficiency will not work. How would you handle this situation?

15. Assume that you as a Firefighter have been given an assignment by your supervisor, to be completed within a specified time frame. What action would you take if after starting the assignment you realize that you are not going to meet the time frame?

16. Assume that you as a Firefighter on a truck company have been asked a technical question by your Captain and you do not know the answer. What action would you take?

17. You, as a Firefighter are confronted with animosity by a particular crew member, what action would you take?

18. Assume that you as a Firefighter are conducting a classroom drill and you want the class to retain a particular essential idea. How would you help them to remember?

19. As a candidate for the position of Firefighter, why do you think it is a good idea for a Fire Captain to delegate assignments?

20. As a Firefighter you are assigned a job by your Fire Captain and then you are called away for another task. You leave another Firefighter in charge of completing the first job. When you return the Fire Captain is there and the job is not complete. The Fire Captain wants an explanation. What will be your reply?

21. What would you do on your first shift as a Firefighter?

22. Assume that you are appointed to the position of Firefighter and assigned to a station where you don't know any of the members, and the Fire Captain's son is next below you on the hiring list, The other members of the crew admire and respect the Fire Captain and believe his son should have been appointed out of list order. What problems does this create for you, what would you do about

23. What actions are you going to take as a Firefighter on a shift and you discover that the Fire Engineer on duty has a bottle of liquor in his locker?

24. On your way to a fire you notice a second fire, what would you do?

25. You are a "rookie" Firefighter and the Firefighters on the other shifts are not performing their duties as well as your shift. This situation is making extra work for the Firefighters on your shift. How would you handle this situation?

SUMMARY

In this chapter we have analyzed the oral interview process. As we look at each component it becomes evident that much work is required in order to be successful enough to obtain a Fire Service career. The better prepared each candidate the better he/she will do.

Preparation is the key to success. It is necessary to learn as much about the profession in general and specific information about the Fire Department that the candidate is testing with. This show interest and lends credibility to each candidates clam as to how serious he/she is about becoming a Firefighter.

Remember the components of the oral interview: Introduction, Background Assessment,, Reasoning/Logic, and Closing Statement. Reviewing possible questions associated with the components will be of great assistance prior to an entrance level Firefighters oral interview.

Along with practice oral interviews there are several other means to improve interview scores. You can develop a list of questions and possible answers and enter them on three (3) by five (5) cards. You would have the questions on one side and the outlined answer on the other side. One advantage of these cards is that you can carry them in your shirt or pant pocket. This will allow for a review whenever you have any spare time on your hands.

Another method to improve your oral skills is to practice in front of a family member or friend. People that have a close relationship with you will probably be more inclined to offer constructive criticism than people that are less familiar to you.

Scheduling practice interviews with Fire Service members will help you develop your oral interview skills. All Firefighters have been through the process at least once. Their knowledge of the types of questions and the expected answers will add to your already increased knowledge of the process.

Now that you have had the opportunity to go over oral interview questions, you should be more prepared for the interview process. It will be your responsibility to improve yourself and:

PRACTICE !   **PRACTICE !**   **PRACTICE !**

ASSIGNMENT:

1. Review Oral Interview questions pages 164 - 173.

CHAPTER 5

MATH CONCEPTS AND ARITHMETIC

MATH CONCEPTS - ARITHMETIC

INTRODUCTION:

The Math concepts and Arithmetic areas that are usually covered on Entrance level Firefighter exams include:

1. Addition.
2. Subtraction.
3. Multiplication.
4. Division.
5. Percentages.
6. Decimals.
7. Fractions.
8. Word Problems.

When encountering the math portion of the exam: work fast but don't concede precision for swiftness. Don't jump to conclusions when computing the answers. Remember when you work out problems in your head it is easy to make a mistake.

Remain calm, even if the problem looks complicated at first. You have seen these types of problems before. Don't panic it will come back to you. If you cannot answer a question, go on to the next one and return to the difficult ones later if you have the time.

The ability to understand math concepts, to solve arithmetical, proportion, and simple algebraic problems is a necessity when taking Fire Service general aptitude examination. Questions involving numerical calculations appear on almost every multiple-choice examination. The type of problems involved are not far removed from those which are found at high school level math classes.

Most of the math questions found on a Firefighter entrance examination can be solved by simple math computation. However, knowledge of proportion and elementary algebra is helpful.

Understanding math concepts is easy for some, but difficult for many candidates. Those of you that are generally weak in arithmetic are urged to brush up on your mathematic skills.

BASIC MATH

Addition, subtraction, multiplication and division of whole numbers is considered elementary arithmetic. Understand how to do these functions should not be a problem for anyone trying to become a Firefighter. However, it is essential that these skills be practiced in order to remain sharp in these basic skills.

FRACTIONS

A fraction is a part of a whole. It is an indicated division: the fraction A/B is equivalent to the division A divided by B. A and B are called the terms of the fraction. The top number in the fraction is called the numerator and the bottom number is called the denominator.

When the numerator of a fraction is less than the denominator, the fraction is considered as part of a whole. The denominator indicates into how many equal parts the whole has been divided, and the nominator indicates how many of those equal parts are being considered.

A proper fraction is a fraction whose numerator is less than its denominator.

An improper fraction is a fraction whose numerator is greater than or equal to its denominator. Any improper fraction has a value greater than or equal to one.

TO MULTIPLY TWO FRACTIONS:

To multiply two fractions, the product of the two fractions equals the product of the numerators divided by the product of the denominators, example:

$$\frac{a}{b} \times \frac{c}{d} = \frac{ac}{bd}$$

TO DIVIDE FRACTIONS:

To divide fractions, the second term is inverted, example:

A/B divided by C/D = A/B X D/C

TO ADD FRACTIONS:

To add fractions it is necessary to change both fractions to equivalent fractions with a common denominator. The least common denominator (LCD) of two or more fractions is the smallest number that is exactly divisible by each of the denominators. Once the LCD has been found, convert each fraction to an equivalent fraction that has the LCD as its denominator. Then, add the like fractions. The last step is to reduce the resulting fraction to its lowest terms. Example: add 1/4 + 1/3 :

$$1/4 = 3/12$$
$$1/3 = 4/12$$
$$3/12 + 4/12 = 7/12$$

TO SUBTRACT FRACTIONS:

To subtract unlike fractions, find the LCD. Convert each fraction to an equivalent fraction that has the LCD as its denominator. Perform the subtraction. Reduce the resulting fraction to lowest terms. Example: subtract 7/8 - 1/4 :

$$7/8 = 7/8$$
$$1/4 = 2/8$$
$$7/8 - 2/8 = 5/8$$

DECIMAL NUMBERS:

A decimal number is a fraction whose denominator is a power of 10. A decimal fraction can be written either in fractional form or with a decimal point. When a number is written in decimal form, the first digit to the right of the decimal point represents tenths, the second digit to the right of the decimal point represents hundredths, and the third digit to the right of the decimal point represents thousandths, and so on.

TO ADD DECIMALS:

To add or subtract decimal numbers, write the numbers under one another with the decimal point in the same vertical line. Add or subtract the columns of like terms just as you would add or subtract whole numbers. Place the decimal point, in the sum or remainder, in the in the same vertical line as the other decimal points. Examples:

ADD:

3.142 + 4.1 + 7.8661

```
  3.142
  4.4
+ 7.8661
 15.1081
```

SUBTRACT:

7.8861 from 10

```
 10.0000
- 7.8862
  2.1139
```

TO MULTIPLY DECIMAL NUMBERS:

To multiply decimal numbers, multiply the numbers just as you multiply whole numbers. Find the number of decimal places in each of the numbers being multiplied, and add these two numbers together. Place the decimal point in the answer so the answer has as many decimal places as the sum found in the previous step. Example, multiply .25 by .35 :

```
      .25
    x.35
     115
     75
    .0865
```

TO DIVIDE DECIMAL NUMBERS:

To divide decimal numbers, take the number you are dividing by, the divisor, and move the decimal point to the right to make a whole number. For example if you are dividing by 7.07, you move the decimal point over so it is 707., but to balance this equation off, you move the decimal place the same distance to the right in the dividend, the number you are dividing into. In any case, whether you move the decimal point in the divisor or not, the decimal point in the quotient (the answer) goes right above the one in the dividend. Example, divide 7.07 into 279.875:

$$7.07 \overline{) 279.875}$$

$$707 \overline{) 27987.5} = 39.6 \text{ (rounded off)}$$

DEFINITION OF PERCENTAGE:

Percent means "per hundred". The symbol for percent is %. Therefore, 7% means: 7/100, 45% means: 45/100, 400% means: 400/100, and so on.

A method of expressing a decimal number in which the amount is always in hundredths defines percent. For example 6/100 is the same as .06, which is the same as 6%.

In order to find what percent a certain number is of another number, it is necessary to isolate the unknown. Example, how would you find what % 15 is of 45. X will be the unknown percent. Remember that what you do to one side of the equation must be done to the other side of the equation.

15 = what percent of 45? (use X for the unknown number)

15 = X % of 45 therefore 15/45 = 1/3

1/3 = X therefore: X = 33 1/3%

Remember when you are trying to find how much a certain per cent of a number is, you must change the percent to a decimal number. For example if you want to find out how much 6% is of 30, before you multiply you must change 6% to a decimal. 6% changes to .06.

Change 6% to .06 and multiply 30 by this number:

.06 X 30 = 1.8

1.8 = 6% Of 30.

DEFINITION OF A RATIO:

A ratio is the relation of like numbers to each other.

The ratio idea enters into all measurements; Such as a pumper, pumper "A", which is a 1,000 GPM pumper and another pumper, pumper "B", which is a 500 GPM pumper. The ratio between pumper "A" and pumper "B" is: 2 to 1. (2 : 1). The ratio between pumper "B" and pumper "A" is 1 to 2 (1 : 2). A simple matter of how the problem is worded.

DETERMINING RATIOS:

The ratio was determined by dividing the first number (1,000) by the second number (500). It is a simple mater of how the problem is worded.

DEFINITION OF PROPORTION:

Proportion is the quality of a ratio. When two ratios are equal, they are said to form a proportion. Usually written:

A : B as C : D or A/B = C/D; The word "as" means equal.

PROPORTION TERMS AND NAMES:

The four terms are 1,2,3, and 4. The two names are extremes and means.

RULES THAT GOVERN PROPORTION:

The following are the rules that govern proportions:

1. The 1st and 4th terms are known as the extremes.

2. The 2nd and 3rd terms are known as the means.

3. The product of the extremes equals the product of the means.

4. The product of the two means equals the product of the two extremes.

5. The product of the two extremes divided by either means gives the other mean.

6. The product of the mean divided by either extreme gives the other extreme.

THE USE OF PROPORTIONS IN SOLVING PROBLEMS

PROBLEM

If a fire truck traveled 40 miles in 2 hours, how far will it travel in 24 hours?

SOLUTION

In this problem we have two factors: miles (distance) and hours (time). The two known quantities of the same subject are 2 hours and 24 hours. The problem ask the question, how far? Thus, the unknown quantity will be X miles (X representing the unknown). There is one other point to consider before solving the problem and that is the obvious fact that the fire truck will travel further in 24 hours than it will in 2 hours. Thus, X is going to be greater than 40.

To simplify proportional problems, remember this simple saying and how it is applied. "The least divided by the most of the same subject is equal to the least divided by the most of the other subject." Now apply the previous statement to the problem:

$$\frac{\text{(least hours) } 2}{\text{(most hours) } 24} = \frac{\text{(least miles) } 40}{\text{(most miles) } X}$$

HOURS — MILES

To complete the above problem, cross multiply and solve for X =

$$2X = 960 \text{ miles.}$$
$$X = 480 \text{ miles.}$$

Cross multiplication is characteristic to proportional problems.

Cross cancellation is not done in proportional problems.

Cancellation in proportional problems form a square.

PROPORTION PROBLEMS

1. If 35 articles of merchandise may be purchased for 25 dollars, how many of the same kind of articles may be purchased for 65 dollars?

2. If the radius of a circle is 50 feet and has a circumference of 100 feet, what will be the circumference of a circle if its radius is 28 feet?

3. If 12 workmen complete a piece of work in 8 days, find the time required by 7 men to complete the same piece of work

4. If a fire truck traveling 35 miles per hour covers a distance in 7 hours, how long will it take a car traveling 50 miles per hour to cover the same distance?

5. If 2 hose lines, of the same diameter, fill a tank in 2 1/2 hours how long will it take for 7 hose lines, each having the same diameter as the first two lines, to fill the same tank?

6. If a vertical rod 6 feet in height will cast a shadow 4 feet long, and a tree casts a shadow 48 feet long, how high is the tree?

TYPES OF PROPORTION

Two types of proportion are "Direct Proportion" and "Indirect Proportion".

An example of Direct Proportion problem: If 3 dozen eggs cost 12 dollars, how much will 6 dozen eggs cost?

In this problem there are two factors: eggs and money. This problem is a Direct Proportion problem because both of these items, eggs and money, go up.

An example of an Indirect Proportion problem: If three men build a fence in 12 days, how many days will it take 6 men to build the same fence?

In this problem there are two factors: days and men. This problem is an Indirect proportion problem because one item (men) goes up and the other item (days) goes down.

SUMMARY

Many of the math problems found on a Firefighter entrance examination can be solved by using concepts and principles we discussed here in this session. It is important that you keep up your math skills. If you don't use it you will loose it. Practice as often as you can and resist the temptation to use a calculator. It is easier to use a calculator rather than your brain. Unfortunately, calculators are not allowed to be used when taking entrance examinations. maintaining basic math skills is like anything else we have discussed in this class. The more you practice the higher your skill level. The higher your skill level the better chance you have to do well on exams.

ASSIGNMENT:

Complete basic math and the basic math problems on pages 185 - 196.

MATHEMATICS WORK SHEET #1

ADD:

1. 482 + 926 + 58 + 389 =

2. 2,713 + 8,649 + 3,574 + 2,020

3. 30.4 + 6.5 =

4. 20 + .75 + 4 + 1.25 =

5. $58.87 + $76.73 + 137.59 =

6. 1/2 + 1/2 =

7. 3/5 + 1/5 =

8. 13 + 2 3/4 =

9. 2 5/8 + 3 3/4 =

SUBTRACT:

10. 5,681 - 796 =

11. 76.8 from 462.53 =

12. 48.3 - 4.8 =

13. 2/3 - 5/8 =

14. 15 2/3 - 12 3/4 =

MULTIPLY:

15. 3,057 X 6 =

16. 7.45 X 9 =

17. $13.30 X 12 =

18. 1/5 X 1/2 =

19. 1 2/5 X 1/3 =

DIVIDE:

20. 427 divided by 7 =

21. 53 divided by 5 =

22. 487.2 divided by 8 =

23. 6 divided by 3 3/5 =

24. 1/5 divided by 1/5 =

25. 4 divided by 1/2 =

ANSWERS
WORK SHEET #1

1. 1,855
2. 16,956
3. 36.9
4. 26
5. $273.19
6. 1
7. 4/5
8. 15 3/4
9. 6 3/8
10. 4,885
11. 385.73
12. 43.5
13. 1/24
14. 2 11/12
15. 18,342
16. 67.05
17. $159.60
18. 1/10
19. 7/15
20. 61
21. 10.6
22. 60.9
23. 1 2/3
24. 1
25. 8

MATHEMATICS WORKSHEET #2

ADDITION PROBLEMS

ADD:

1. 42 + 33 + 18 =

    A. 83		B. 93		C. 91		D. 79		E. None

2. 18 + 16 + 9 =

    A. 39		B. 43		C. 49		D. 53		E. None

3. 53 + 26 + 17 =

    A. 86		B. 89		C. 93		D. 96		E. None

4. 5 + 13 + 88 + 102 =

    A. 198		B. 208		C. 193		D. 203		E. None

5. 58 + 62 + 104 + 153 =

    A. 348		B. 357		C. 376		D. 377		E. None

6. .55 + .34 =

    A. .86		B. .87		C. .89		D. .90		E. None

7. .96 + .88 =

    A. 1.88		B. 1.86		C. 1.84		D. 1.80		E. None

8. 1.10 + .99 =

   A. 1.99   B. 2.09   C. 2.19   D. 2.28   E. None

9. 21.9 + 2.3 =

   A. 23.3   B. 23.9   C. 24.2   D. 24.4   E. None

10. 88.8 + 99.9 =

    A. 178.8   B. 187.8   C. 188.7   D. 188   E. None

11. 3/4 + 3/4 =

    A. 1   B. 1 1/4   C. 1 1/2   D. 1 1/34   E. None

12. 1/2 + 1/4 =

    A. 3/4   B. 1   C. 1 1/4   D. 1 1/2   E. None

13. 1/3 + 1 1/3 =

    A. 2/3   B. 1   C. 1 1/2   D. 1 2/3   E. None

14. 5/8 + 7/8 =

    A. 1   B. 1 1/8   C. 1 1/2   D. 1 5/8   E. None

15. 1/5 + 1/10 =

    A. 2/5   B. 3/10   C. 4/5   D. 1   E. None

SUBTRACTION PROBLEMS

SUBTRACT:

1. 93 - 64 =

   A. 28   B. 29   C. 31   D. 38   E. None

2. 48 - 16 =

   A. 28   B. 29   C. 32   D. 33   E. None

3. 77 - 65 =

   A. 12   B. 14   C. 15   D. 16   E. none

4. 98 - 29 =

   A. 66   B. 68   C. 73   D. 74   E. None

5. 142 - 133 =

   A. 7   B. 8   C. 9   D. 10   E. None

6. .55 - .46 =

   A. .06   B. .07   C. .08   D. .09   E. None

7. .99 - .79 =

   A. 20   B. 2.0   C. .22   D. .20   E. None

8. 1.08 - .76 =

   A. .23   B. .32   C. .33   D. .34   E. None

9. 22.64 - 22.46 =

    A. .18   B. 1.8   C. 18   D. 1.3   E. None

10. 132.1 - 113.9 =

    A. 18   B. 18.2   C. 12.8   D. 12   E. None

11. 3/4 - 3/4 =

    A. 1/4   B. 1/3   C. 1/8   D 1/10   E. None

12. 3/4 - 1/4 =

    A. 1/4   B. 1/3   C. 1/2   D. 3/4   E. None

13. 1/3 - 1/4 =

    A. 1/8   B. 1/6   C. 1/10   D. 1/12   E. None

14. 1 1/4 - 1 1/8 =

    A. 1/4   B. 1/8   C. 1/3   D 1/6   E. None

15. 22 1/2 - 15 1/6 =

    A. 7   B. 7 1/6   C. 7 1/2   D. 7 1/3

MULTIPLICATION PROBLEMS

MULTIPLY:

1. 3 X 87 =

   A. 251   B. 255   C. 261   D. 266   E. None

2. 6 X 58 =

   A. 348   B. 358   C. 361   D. 364   E. None

3. 9 X 98 =

   A. 872   B. 874   C. 878   D. 882   E. None

4. 21 X 44 =

   A. 904   B. 914   C. 924   D. 925   E. None

5. 18 X 111 =

   A. 1998   B. 1999   C. 2008   D. 2011   E. None

6. .25 x .25 =

   A. .0625   B. .625   C .602   D. .620   E. None

7. .5 X .5 =

   A. .10   B. .20   C. .25   D. .30   E. None

8. .325 x .20 =

   A. 6.5   B. .65   C. .065   D. .0065   E. None

9. 1.5 X 1.5 =

    A. 2.05   B. 2.25   C. 2.50   D. 2.55   E. None

10. .1 X .1 =

    A. 1   B. .1   C. .01   D. .11   E. None

11. 1/4 X 1/4 =

    A. 1/8   B. 1/16   C. 1/24   D. 1/32   E. None

12. 1/8 X 1/4 =

    A. 1/8   B. 1/16   C. 1/24   D. 1/32   E. None

13. 3/4 X 3/4 =

    A. 1/2   B. 9/32   C. 9/16   D. 3/4   E. None

14. 7/8 X 1/4 =

    A. 7/8   B. 7/16   C. 7/32   D. 7/64   E. None

15. 5/8 X 4/5 =

    A. 3/8   B. 1/2   C. 5/8   D. 3/4   E. None

DIVISION PROBLEMS

DIVIDE:

1. 64 by 8 =

   A. 7　　　B. 8　　　C. 9　　　D. 10　　　E. None

2. 55 by 5 =

   A. 8　　　B. 9　　　C. 10　　　D. 11　　　E. None

3. 48 by 6 =

   A. 6　　　B. 8　　　C. 9　　　D. 10　　　E. None

4. 42 by 7 =

   A. 4　　　B. 5　　　C. 6　　　D. 8　　　E. None

5. 108 by 12 =

   A. 9　　　B. 12　　　C. 15　　　D. 17　　　E. None

6. 6.4 by 3.2 =

   A. 1.5　　　B. 1.6　　　C. 1.8　　　D. 2　　　E. None

7. 3.3 by 1 =

   A. 1.3　　　B. 3.1　　　C. 3　　　D. 3.3　　　E. None

8. 14.44 by 7.22 =

   A. 7.22　　　B. 7　　　C. 3.22　　　D. 3　　　E. None

9. .444 by 2.2 =

   A. 2.02   B. 20.2   C. .202   D. .020   E. None

10. .15 by .1 =

    A. 15   B. 1.5   C. .15   D. .015   E. None

11. 3/4 by 3/4 =

    A. 1/2   B. 3/4   C. 1   D. 1 1/2   E. None

12. 3/4 by 2/3 =

    A. 1 1/6   B. 1 1/3   C. 1 1/8   D. 1   E. None

13. 1/8 by 1/16 =

    A. 2/16   B. 1/8   C. 1   D. 2   E. None

14. 7/8 by 1/16 =

    A. 11   B. 1.1   C. 14   D. 1.4   E. None

15. 1/2 by 1/3 =

    A. 1 1/2   B. 1 1/3   C. 1   D. 7/8   E. None

PERCENTAGE PROBLEMS

PERCENT =

1. 10% of 30 =

   A. 3　　B. 3.3　　C. 10　　D. 20　　E. None

2. 15% of 60 =

   A. 7　　B. 8　　C. 9　　D. 13　　E. None

3. 25% of 300 =

   A. 50　　B. 55　　C. 65　　D. 75　　E. None

4. 20% of 30 =

   A. 6　　B. 15　　C. 3　　D. 10　　E. None

5. 75% of 96 =

   A. 52　　B. 57　　C. 77　　D. 82　　E. None

6. 9 is what per cent of 12 ?

   A. 3/4%　　B. 7.5%　　C. 75%　　D. 25%　　E. None

7. 10 is what per cent of 30 ?

   A. 2/3%　　B. 33%　　C. 1/3%　　D. 15%　　E. None

8. 6 is what per cent of 60 ?

   A. 8%　　B. 80%　　C. 10%　　D. 1.1%　　E. None

9. 10 is what per cent of 20 ?

   A. 10%   B. 20 %   C. 45%   D. 50 %

10. 90 is what per cent of 200 ?

    A. 90%   B. 99%   C. 45%   D. 49%   E. None

11. 100 is what per cent of 300 ?

    A. 33%   B. 34%   C. 35%   D. 36%   E. None

12. 99 is what per cent of 100 ?

    A. 1%   B. 9%   C 90 %   D. 99%   E. None

13. 46 is what per cent of 200 ?

    A. 23%   B. 46%   C. 92%   D. 50%   E. None

14. 50 is what per cent of 200 ?

    A. 50%   B. 40%   C. 25%   D. 15%   E. None

15. 80 is what per cent of 400 ?

    A. 60%   B. 80%   C. 20%   D. 40%   E. None

WORD PROBLEMS

1. The money spent by Captain Sutton for class room study was as follows: $2.00 for paper; $2.29 for notebook; $1.00 for pencils; 99 cents for a ruler; 29 cents for an eraser and 39 cents for sales tax. What was his total bill?

    A. $6.39
    B. $6.46
    C. $6.69
    D. $6.96

2. Fireman Simpson works in a machine shop on his off duty days. During one month he made the following number of machined parts: 598; 699; 750; 433; 1,200. how many parts did he make?

    A. 3,633.
    B. 3,680.
    C. 3,733.
    D. 3,780.

3. Firefighter Simpson went to the market with a $20 bill. He bought the following items: roast, $7.87; potatoes, $2.34; milk, $1.88; salad fixings $4.65; ice cream, $2.69. How much change did he bring back to the fire station?

    A. 57 cents.
    B. 61 cents.
    C. 66 cents.
    D. 87 cents.

4. The City Fire Departments area is 298 square miles. The County Fire Departments area is 2,342 square miles. How much larger is the County Fire Department area?

    A. 1,942 square miles.
    B. 2,034 square miles.
    C. 2,044 square miles.
    D. 2,054 square miles.

5. The City Fire Department can buy the fire apparatus for $210,354 if it pays cash. If the City Fire Department takes 12 months to pay, it will cost $223,989. How much will they save by paying cash?

    A. $13,365.
    B. $13,635.
    C. $13,656.
    D. $13,665.

6. The year 1991 is the 68th anniversary of the City Fire Department. When was the City Fire Department founded?

    A. 1867.
    B. 1888.
    C. 1923.
    D. 1926.

7. The Firefighter Association's telephone bill listed the following items: $ 87.50 service for long distance; 890 message units at 5 cents each; 22.34 service charge. What was the total bill?

    A. $134.54.
    B. $143.34.
    C. $154.34.
    D. $154.43.

8. Engine 82 is going to the mechanic for re-power. If the trip covers 644 miles round trip and the engine can get 7 miles per gallon at a cost of $1.39 per gallon for fuel. What will be the cost of the fuel for the trip?

    A. $127.88.
    B. $128.87.
    C. $137.88.
    D. $148.87.

9. Engine 88 uses about 6 gallons of fuel for each mile. If it has been fueled with 50 gallons, approximately how far can it go?

    A. 200 miles.
    B. 250 miles.
    C. 300 miles.
    D. 350 miles.

10. A strike team leaves for a brush fire at 2115 hours and arrives at the scene at 2445 hours. The strike team averaged 31 miles per hour. How many miles did they travel?

    A. 105.8 miles.
    B. 108.0 miles.
    C. 108.5 miles.
    D. 118.5 miles.

11. During three 24 hour shift periods: E-81 traveled 66 miles, E-82 traveled 88 miles, E-83 traveled 75 miles, T-81 traveled 33 miles, T-82 traveled 44 miles, R-81 traveled 106 miles, R-82 traveled 155 miles, and C-81 traveled 64 miles. How much more or less than 1,000 miles did there combined total equal?

    A. 369 miles more.
    B. 369 miles less.
    C. 639 miles more.
    D. 639 miles less.

12. Engineer Harper has four lengths of 2 1/2" hose in the following lengths: 49 feet, 50 feet, 48 feet, and 47 feet. What is the total length of these pieces?

    A. 147 feet.
    B. 149 feet.
    C. 194 feet.
    D. 197 feet.

13. During one year the City Fire Departments HQ station had 5,566 total responses. During the same period Station #2 had 6,434 total responses. How many more responses did Station #2 have than HQ?

    A. 686.
    B. 688.
    C. 866
    D. 868.

14. During the first half of the year, the City Fire Department spent 75 per cent of its $1,000,000.00 budget. What is the total amount of monies available from the budget for the remainder of the year?

    A. $250,000.00.
    B. $275,000.00.
    C. $300,000.00.
    D. $325,000.00.

15. Firefighter Ramey makes $1,500.00 per pay period. He saves 6 per cent of his earnings each pay period. How much does he save each pay period?

    A. $66.00.
    B. $69.00.
    C. $90.00.
    D. $96.00.

16. The City Fire Department last year required each hazardous business owner to pay $.0088 of their profits for permits to do business within the City. If the profit of a business was a total of $15,000.00, how much did they pay for a permit?

    A. $103.00.
    B. $113.00.
    C. $122.00.
    D. $132.00.

17. B/C Pedego paid $.75 per pound for "Spill-Kill". He purchased 12 bags and each bag weighed 25 pounds. What was the total cost?

    A. $200.00.
    B. $205.00.
    C. $210.00.
    D. $225.00.

18. Firefighter Simpson called in sick with a temperature of 101.3 degrees. If the normal temperature is 98.6 degrees, how many degrees above normal was he?

    A. 2.2 degrees.
    B. 2.5 degrees.
    C. 2.7 degrees.
    D. 3.3 degrees.

19. What will be the cost of 300 feet of fire hose at a cost of $1.457 per foot?

    A. $437.10.
    B. $431.70.
    C. $423.10.
    D. $418.70.

20. In three successive shifts, "A" shift spent $34.19, $28.16, and $42.22 for meals. What was the total cost for the meals during this period?

    A. $103.17.
    B. $105.17.
    C. $113.07.
    D. $117.03

21. Firefighter Wilkinson bought a pen, paper, and envelopes. The entire bill was $7.50. The pen cost $2.39 and the paper was $3.25. What was the cost of the envelopes?

    A. $1.36.
    B. $1.56.
    C. $1.66.
    D. $1.86.

22. In a promotional exam, Engineer Ortega answered 95% of the questions correctly. There were 140 questions. How many questions did he get right?

    A. 139.
    B. 133.
    C. 127.
    D. 123.

23. The City Fire Departments baseball team played 20 games last season and lost 25% of them. How many games did they win?

    A. 5 games.
    B. 10 games.
    C. 15 games.
    D. 17 games.

24. In the first promotional exam that Captain Reardon took there were 100 questions and he had 80 questions correct. In the next exam having the same number of questions he had 10% more correct than he did in the first exam. How many did he have correct on the second exam?

    A. 88.
    B. 90.
    C. 92.
    D. None of the above.

25. Fireman Audet bought a flashlight for $7.50. After using it for two shifts he sold it for a 20% loss. At what price did he sell the flashlight?

    A. $5.00.
    B. $5.50.
    C. $6.00.
    D. $6.50.

26. The Firefighter Auxiliary earned a $120.00 commission from the sale of magazine subscriptions. This amount was equal to 25% commission on all sales. How many dollars worth of magazine subscriptions did they sell?

    A. $240.00.
    B. $480.00.
    C. $340.00.
    D. $680.00.

27. The State Forestry Division of Fire map has a scale where 1 inch = 100 miles. How many miles are represented by 2 1/4 inches on the map?

    A. 225 miles.
    B. 220 miles.
    C. 250 miles.
    D. 200 miles.

28. Fireman Hoyem's salary was increased from $1,000.00 per pay period to $1,200.00 per pay period. What per cent of increase is this?

    A. 5% increase.
    B. 10% increase.
    C. 15% increase.
    D. 20% increase.

29. Last year the City Fire Department responded to 8,875 incidents. This year they responded to 9,230 incidents. What per cent of increase is this?

    A. 4%.
    B. 5%.
    C. 6%.
    D. None of the above.

30. Fireman Paramedic Cannon wanted an early relief from Fireman Paramedic Roberts at 0500 hours. Roberts was 1 hour and 40 minutes late. At what time did he arrive?

    A. 0540 hours.
    B. 0630 hours.
    C. 0634 hours.
    D. None of the above.

31. Firefighter Simpson is 5 feet, 9 inches tall and Captain Sutton is 67 inches tall. Which man is taller.

    A. Firefighter Simpson.
    B. Captain Sutton.
    C. They are the same height.

32. How much would a Firefighter pay for 5 "Pen-Lights" at a rate of $6.00 per dozen?

    A. $1.50.
    B. $2.00.
    C. $2.50.
    D. $3.00.

33. If 14% of the Firefighters live within the City, what per cent live outside of the City.

    A. 86%.
    B. .14%.
    C. .86%.
    D. None of the above.

34. Firefighter Smith had three Library books that were 4 days overdue. His fine was $.12 per day for each book. How much was his total fine?

    A. $1.24.
    B. $1.34.
    C. $1.43.
    D. $1.44.

35. A year's subscription of "Firefighters" monthly magazine cost $19.95 per year. If a single copy cost $2.00, how much will a Firefighter save by taking a year's subscription instead of buying a single copy each month?

    A. $3.05.
    B. $4.05.
    C. $5.03.
    D. $5.04.

36. What is the difference in cost to a firefighter between a flashlight listed at $50.00 less 10% and one listed at $45.00 less 20% ?

    A. $4.00.
    B. $4.50.
    C. $2.00.
    D. $2.50.

37. Firefighter Wendl borrowed $4000.00 at 14%. How much additional money can he borrow at 16% if his total interest is not to exceed $1250.00 a year?

    A. $ 800.00.
    B. $ 900.00.
    C. $2000.00.
    D. $4000.00.

38. Fire Engineer Hibben wishes to construct a storage box 12 feet long to keep spare equipment in. If each piece of equipment requires 4 square feet of space, how wide should he construct the box ?

    A. 6 1/2 feet.
    B. 6 feet 8 inches.
    C. 20 feet.
    D. 80 feet

39. How much longer does it take a ladder truck to travel one mile at 20 miles per hour than at 30 miles per hour?

    A. 1 minute.
    B. 10 minutes.
    C. 20 minutes.
    D. 40 minutes.

40. Fire Captain Hyatt stated to Firefighter Murray that light travels approximately 186,000 miles a second and that the sun is 933 million miles away from the earth. Firefighter Murray asked Captain Hyatt if he could figure out how long it would take a ray of light to travel from the sun to the earth. Captain Hyatt said that he could figure it out to the nearest minute. What would be the correct time?

    A. 2 minutes.
    B. 5 minutes.
    C. 8 minutes.
    D. 58 minutes.
    E. None of the above.

ANSWER SHEET

MATH PROBLEMS

ADDITION	SUBTRACTION	MULTIPLICATION	DIVISION	PERCENTAGE
1. B	1. B	1. C	1. B	1. A
2. B	2. C	2. A	2. D	2. C
3. D	3. A	3. D	3. B	3. D
4. B	4. E	4. C	4. C	4. A
5. D	5. C	5. A	5. A	5. E
6. C	6. D	6. A	6. D	6. C
7. C	7. D	7. C	7. D	7. B
8. B	8. B	8. C	8. E	8. C
9. C	9. A	9. B	9. C	9. D
10. C	10. B	10. C	10. B	10. C
11. C	11. E	11. B	11. C	11. A
12. A	12. C	12. D	12. C	12. D
13. D	13. D	13. C	13. D	13. A
14. C	14. B	14. C	14. C	14. C
15. B	15. D	15. B	15. A	15. C

WORD PROBLEMS

1. D	11. C	21. D	31. A
2. B	12. C	22. B	32. C
3. A	13. D	23. C	33. A
4. C	14. A	24. A	34. D
5. B	15. C	25. C	35. B
6. C	16. D	26. B	36. A
7. C	17. D	27. A	37. A
8. A	18. C	28. D	38. B
9. C	19. A	29. A	39. A
10. C	20. B	30. D	40. C

CHAPTER

6

AREAS
VOLUMES
CAPACITIES
AND
BASIC HYDRAULICS

FORMULAS AND CONVERSIONS

INTRODUCTION

In this chapter we will identify basic formulas to find areas, volumes and capacities. We will also discuss heat measurements and basic hydraulics. The information presented here is important because it is closely associated to what Firefighters must know when dealing with hydraulics.

This information will not only help you get on the job, it will also help you when you are on the job.

MEASUREMENT OF AREAS, VOLUMES, AND CAPACITIES

1. SQUARES EXPRESSED IN SQUARE UNITS:

    A. A figure having parallel sides and four right angles.

    B. The formula for finding the ares of a square is:

    $$AREA = L \times W$$
    or
    $$A = \text{the square of the sides: } A = (S)^2 ; \text{ or } A = (S)(S)$$

2. RECTANGLES EXPRESSED IN SQUARE UNITS:

    A. A figure having parallel sides and four right angles.

    B. The formula for finding the area of a rectangle is:

    $$AREA = L \times W$$

3. CIRCLES EXPRESSED IN SQUARE UNITS:

    A. A plane bounded by a curved line, every point of the curve line is the same distance from the center.

    B. The formula for finding the area of a circle is:

    $$AREA = .7854 \times (D)^2$$

    OR

    AREA = "pie" times the square of the radius: expressed as: "pie" r square which is: $\pi (r)^2$
    ("pie") π = 3.1416 or 22/7

    C. The distance from the center to the curved line is called the radius (r). The formula for finding the radius of a circle is:

    $$r = \sqrt{\frac{AREA}{3.1416}}$$

209

D. The distance from curved line to curved line through the center is called the diameter (D). The formula for finding the diameter of a circle is:

$$D = \text{the circumference divided by "pie" } (\pi): \frac{C}{3.1416}$$

OR

$$D = \text{the square root of the area divided by .7854: } \frac{\sqrt{A}}{.7854}$$

E. Circumference is the total distance around the outside of a circle (C). The formula for finding the circumference of a circle is:

$$C = \text{"pie" times the diameter: } \pi D$$

4. VOLUME OR CAPACITY EXPRESSED IN CUBIC UNITS:

A. To find the volume of a container, we must consider a third dimension: height or depth.

B. The formula for finding volume is:

$$\text{VOLUME} = \text{AREA TIMES HEIGHT: } V = AH$$

C. The formula for finding the volume of rectangles or squares is:

$$\text{VOLUME} = LWH$$

D. The formula for finding the volume of cylinders is:

VOLUME = .7854 times the diameter squared times the height.

$$\text{VOLUME} = .7854\, D^2 H$$

E. To find the number of gallons for a fifty foot section of fire hose square the hose diameter of the hose then multiply by 2.04:

$$\text{GALLONS} = 2.04\, D^2$$

5. AREA OF A TRIANGLE:

A. A figure formed by three lines intersecting by two's in three points, and so forming three angles.

B. The formula for finding the area of a right triangle is:

AREA = AB/2 (altitude times base divided by 2): $\dfrac{AB}{2}$

C. The formulas for finding the length of one side of the right triangle when the other two lengths are known, are:

1. C = the square root of A square plus B square:

$$C = \sqrt{A^2 + B^2}$$

2. A = the square root of C square minus B square:

$$A = \sqrt{C^2 - B^2}$$

3. B = the square root of C square minus A square:

$$B = \sqrt{C^2 - A^2}$$

A = ALTITUDE; B = BASE; C = HYPOTENUSE

RIGHT TRIANGLE

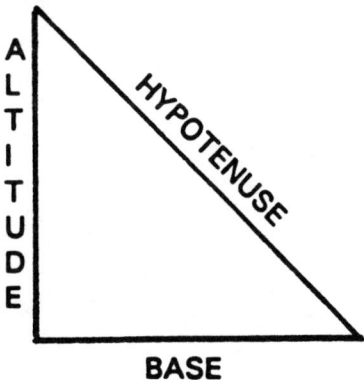

```
A
L
T
I       HYPOTENUSE
T
U
D
E
        BASE
```

6. TEMPERATURE CONVERSIONS:

A. To change Centigrade to Fahrenheit:

$$F = 9/5\ C + 32$$

B. To change Fahrenheit to Centigrade:

$$C = 5/9\ (F - 32)$$

C. EXAMPLE #1: Centigrade = -40 F = ?

$$F = 9/5\ (-40) + 32$$

$$F = (-12) + 32 = -40$$

NOTE: THIS IS THE ONLY TEMPERATURE WHERE CENTIGRADE AND FAHRENHEIT ARE THE SAME!

SOME PRINCIPLES OF FLUID PRESSURE:

1. Liquid pressure is exerted in a perpendicular direction to any surface on which it acts. This principle is represented in figure #1 with a vessel that has flat sides. This vessel contains water. The pressure exerted by the weight of the water is perpendicular to the walls of the container. If this pressure is exerted in any other direction, as indicated by the arrows, the water would start moving downward along the sides and rise in the center.

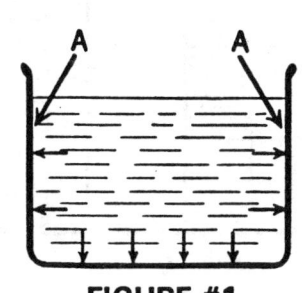

FIGURE #1
Small arrows show direction of water pressure.

2. At any given point beneath the surface of a liquid, the pressure is the same in all directions - downward, upward, and sideways. This principle is represented in figure #2 with a coupling of a hoseline that has two pressure gauges inserted. When water is at rest, while the nozzle is shut down, both gauges will indicate the same pressure, since the pressure at a point in a fluid at rest is of the same intensity in all directions.

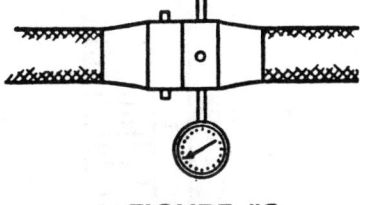

FIGURE #2
Pressure is the same at both gages.

3. When pressure applied to a confined liquid, pressure is transmitted in all directions without reduction in intensity. Figure #3 illustrates this principle with a hollow sphere that has a water pump attached. A series of gauges are set into the sphere around its circumference. With the sphere filled with water and pressure applied by the pump, all gauges will indicate the same reading, provided they are on the same grade line with no change in elevation.

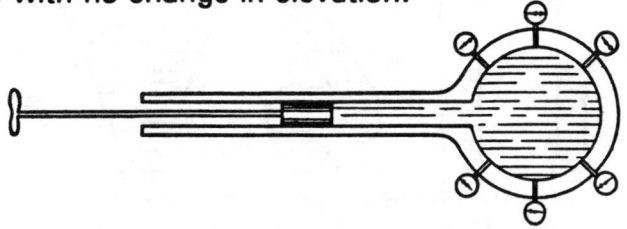

FIGURE #3

4. The pressure of a liquid in an open vessel is proportional to the depth of the liquid. Figure #4 illustrates this principle with the use of three vertical containers each being one square inch in cross-sectional area. The depth of the water is one foot in the first container, two feet in the second container, and three feet in the third container. The pressure at the bottom of the second container is twice that of the first container. Therefore the pressure of a liquid in an open container is proportional to its depth.

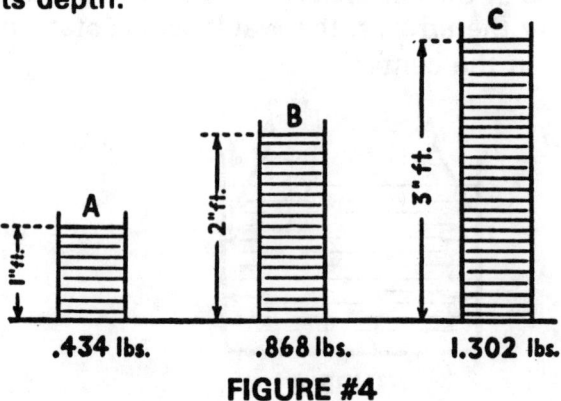

.434 lbs.   .868 lbs.   1.302 lbs.

FIGURE #4

5. The pressure of a liquid in an open vessel is proportional to the density of the liquid. Figure #5 illustrates this principle with the use of two containers, one containing mercury one inch in depth, the other containing water at a depth of 13.55 inches. The pressure at the bottom of each container is approximately the same since mercury is approximately 13.55 times denser than water. Therefore, the pressure of a liquid in an open vessel is proportional to the density of the liquid.

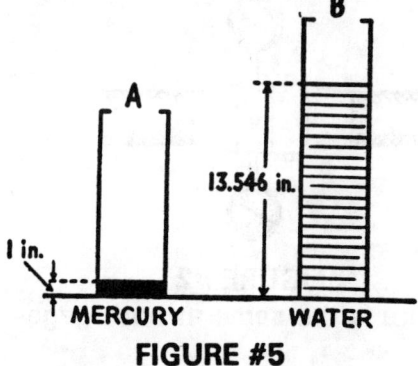

FIGURE #5

6. Liquid pressure on the bottom of a vessel is unaffected by the size and shape of the vessel. Figure #6 illustrates this principle by the use of fpur different containers of four different shapes, each having the same cross-sectional area at the bottom and having the same height.

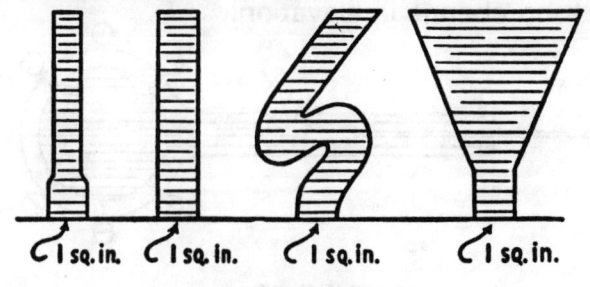

FIGURE #6

ELEMENTARY PRINCIPLES OF HYDRAULICS

1. DEFINITION OF HYDRAULICS:

    A. Hydraulics is a branch of applied science mechanics that deals with the physical characteristics exhibited by fluids at rest and in motion.

    B. Hydrostatics deals with fluids at rest.

    C. Hydrokinetics deals with fluids in motion.

2. WATER CONSUMPTION:

    A. U.S. Cities now produce and distribute approximately 25 billion gallons of water per day.

    B. For every 1,000 persons in the U.S., there is about 3 miles of water main.

    C. The per-capita consumption of water is 150 gallons per day.

    D. The average per-capita use of water for fire protection is 10-15 gallons of water per day.

3. CHARACTERISTICS OF WATER:

    A. Below 32 degrees Fahrenheit water exist as a solid.

    B. From 32 degrees Fahrenheit to 212 degrees Fahrenheit water exist as a liquid.

    C. Above the boiling point of water (212 degrees Fahrenheit), water exists in the form of a gas known as water vapor.

    D. Water has a maximum density @ 39.9 degrees Fahrenheit or 4 degrees Centigrade.

SOME IMPORTANT MEASURES

1. MEASURES OF LENGTH:

12 inches	= 1 foot
3 feet	= 1 yard
5 1/2 yards	= 1 rod
320 rods	= 1 mile
1760 yards	= 1 mile
5280 feet	= 1 mile.

2. SQUARE MEASURE:

144 square inches	= 1 square foot
9 square feet	= 1 square yard
30 1/4 square yards	= 1 square rod
160 rods	= 1 acre
640 acres	= 1 square mile

3. AVOIRDUPOIS WEIGHT:

16 ounces	= 1 pound
100 pounds	= 1 hundredweight
2000 pounds	= 1 ton
2240 pounds	= 1 long ton

4. LIQUID MEASURE:

4 gills	= 1 pint
2 pints	= 1 quart
4 quarts	= 1 gallon
31 1/2 gallons	= 1 barrel
2 barrels	= 1 hogshead

5. TROY WEIGHT: (Gold, silver, etc.)

24 grains	= 1 pennyweight
20 pennyweight	= 1 ounce
12 ounces	= 1 pound

6. DRY MEASURE:

2 pints	= 1 quart
8 quarts	= 1 peck
4 pecks	= 1 bushel

ROMAN NUMERAL EQUIVALENTS

I	= 1		XVI	= 16
II	= 2		XVII	= 17
III	= 3		XVIII	= 18
IV	= 4		XIX	= 19
V	= 5		XX	= 20
VI	= 6		XXI	= 21
VII	= 7		XXX	= 30
VIII	= 8		XL	= 40
IX	= 9		L	= 50
X	= 10		LX	= 60
XI	= 11		LXX	= 70
XII	= 12		LXXX	= 80
XIII	= 13		XC	= 90
XIV	= 14		C	= 100
XV	= 15		CC	= 200

WATER MEASUREMENTS

UNITED STATES GALLONS:

1. One cubic foot of fresh water weighs about 62.5 pounds.

2. One cubic foot of ocean water weighs about 64 pounds.

3. One cubic foot of water contains 1728 cubic inches.

4. One cubic foot of water contains 7.481 U.S. gallons.

5. One U.S. gallon of water contains 231 cubic inches.

6. One U.S. gallon of water contains 128 fluid ounces.

7. One U.S. gallon of fresh water weighs 8.35 pounds.

8. A column of water one inch square and one foot high weighs .434 pounds.

9. A column of water 2.304 feet high will exert a pressure of one pound per square inch at its base.

CONVERSION FACTORS:

MULTIPLY	BY	TO OBTAIN
Cubic feet	7.481	U.S. gallons
Cubic feet	62.5	Pounds fresh water
Cubic feet	64.0	Pounds ocean water
Feet of water	0.8819	Inches of mercury
Feet of water	62.5	Pounds per sq. foot
Feet of water	0.4335	pounds per sq. inch
U.S. gallons	8.35	Pounds
U.S. gallons	128.0	Fluid ounces
U.S. gallons	0.1337	Cubic feet
U.S. gallons	231.0	Cubic inches
U.S. gallons	0.833	Imperial gallons
Inches of mercury	1.133	Feet of water
Inches of mercury	0.4912	Pounds per sq. inch

SUMMARY

Knowledge of the material discussed in this session will assist you in your efforts in obtaining a career in the Fire Service. The basic theories identified should be thoroughly learned, as they are the basis for understanding hydraulics. If you understand all the concepts discussed in this session, you should have no problem getting through math portions of the entrance examination.

ASSIGNMENT:

1. Worksheet: "Measurements and Basic Hydraulics", on pages 220 - 221.

WORKSHEET #1

MEASUREMENTS AND BASIC HYDRAULICS

1. Find the area of a square with one side 7 inches high:

    ANSWER = 49 square inches.

2. Find the area of a rectangle having a length of 10 inches and a width of 5 inches:

    ANSWER = 50 square inches.

3. What would be the area of a circle that has a diameter of 3 inches?

    ANSWER = 7.0686 square inches; rounded off = 7.07 square inches.

4. Find the diameter of a circle that has an area of 9 inches:

    ANSWER = 3.82 inches.

5. What is the circumference of a circle that has a diameter of 7 inches?

    ANSWER = 21. 99 inches; rounded off = 22 inches.

6. Find the volume of a rectangle that has a length of 10 inches, a width of 5 inches and a height of 3 inches:

    ANSWER = 150 inches.

7. What would be the volume, in cubic inches, of a cylinder that has a diameter of 2.5 inches and a height of 50 feet?

    ANSWER = 2,945.25 cubic inches.

8. What would be the length of the altitude of a right triangle that had a hypotenuse (C) of 9 inches, and a base (B) of 4 inches?

   ANSWER = 8.89 inches.

9. Change 40 degrees Centigrade to the equivalent in Fahrenheit:

   ANSWER = 104 degrees Fahrenheit.

10. Change 40 degrees Fahrenheit to the equivalent in Centigrade:

    ANSWER = 4.44 degrees Centigrade.

11. What would be the pressure at the base of a cylinder 5 feet high and having a diameter of 3 inches?

    ANSWER = 2.17 pounds per square inch. (PSI)

12. What would be the pressure at the base of a cylinder 5 feet high and having a diameter of 6 inches?

    ANSWER = 2.17 PSI.

13. How high would a column of water be in a cylinder that exerted a pressure of 3.689 PSI?

    ANSWER = 8.5 feet.

14. A cylinder tank is 10 feet in diameter, 30 feet in height, and is 3/4 full of water. How many gallons of water are in the tank?

    ANSWER = 15,220 gallons.

15. A rectangle tank which is 7 feet 6 inches wide and 22 feet 6 inches long is filled with water to a depth of 10 feet. This tank is supplying a pumper which is discharging 265 gallons per minute (GPM); at this rate, the water supply will last nearly?

    ANSWER = 48 minutes.

CHAPTER 7

READING COMPREHENSION: IMPROVING READING SKILLS

READING COMPREHENSION: IMPROVING READING SKILLS

INTRODUCTION

It makes no difference whether you take a general knowledge examination or a job selected examination, reading comprehension plays an important part in order to be successful.

Entrance Firefighter exams will have reading comprehension portions in order to test your ability to read, understand and retain information that you may be required to read during your performance as a Firefighter.

It has been said "if you can read you can do anything". This statement certainly applies to taking written exams. Another common statement: "he/she is a good test taker". It has been found that the good test takers are always good readers. A person that is a good reader reads fast and retains the material. In this session we will identify methods to increase reading speed and comprehension. The end result will be to make you a better reader and a good test taker.

SOME POINTS TO REMEMBER WHILE TAKING AN EXAM:

1. Read the entire paragraph prior to looking at the questions relating to the paragraph.
2. Sort out the principal view of the reading.
3. Locate distinctive components in the material.
4. Ascertain the significance of peculiar words.
5. Ascertain the distinctive method that the material is written so as to achieve its desired effect.
6. Use any prior knowledge that you have to complement the concept of the material.

Another requirement towards success is to improve your reading skills. It is necessary to read faster and comprehend what is read. This is necessary because there is so much material to read and most people don't have enough time to devote to reading. Because of this fact, it is necessary to learn to read efficiently, with at least a little more speed and a lot more understanding.

It is now time to read with mature proficiency. That means gaining important new skills. It means organizing your present skills and the new ones into a reliable system for mastering various types of reading matter. It means suppressing bad habits. It also means increasing your reading speed somewhat, and increasing very much the flexibility and control with which you use different reading speeds.

EFFICIENT READING:

The following statement is often heard: "He/she is a good test taker". How does a person become a good test taker? The answer is by becoming a good reader. Efficient reading includes reading faster and with more understanding.

The first aspect to consider is the sheer bulk of Fire Service related material there is available to read. Secondly, the subject matter is often difficult. Material from physics, chemistry, hydraulics, fire fighting tactics and strategy, apparatus, tools and equipment is often foreign to entrance level Firefighter candidates. A high degree of reading skill is necessary if you are to comprehend it as thoroughly as you need. Efficient reading includes how to look material over before reading it thoroughly, making a map to guide you through the unknown territory ahead. This important technique is called pre-reading. Efficiency also includes reading faster and comprehending or retaining more information.

READING FASTER:

The question is often asked: When speed goes up, is comprehension reduced? There is a consistent fear among untrained readers that when reading speed is increased, there is a corresponding loss of understanding. This is largely unfounded. On the contrary, most readers find that their comprehension improves as they learn to read faster, reading comprehension improves. The reason for this is reading faster requires discipline, and concentration. Reading is like typing or tennis: a sudden spurt of speed just makes for wildness and mistakes, but with practice and instruction, speed and accuracy both improve, reinforcing each other.

Speed becomes a tool that you control to get the comprehension you require. The faster pace helps keep your attention centered on your reading. Most importantly, to develop higher speeds you must learn to read by phrases rather than word by word.

PRE-READING:

The procedure for pre-reading is as follows:
1. Read normally the first three paragraphs of the chapter.
2. Read the first sentence only of each remaining paragraph.
3. Read the last three paragraphs thoroughly.

If the material you are reading has the proper format and organization it will be comprised of three components. These components are the introduction, body, and the summary. An old saying that indicates the proper writing format is: "first tell them what you are going to tell them, then tell them, then tell them what you told them".

Pre-reading makes comprehension easier and better, because it provides a framework into which you fit details during your later through reading. The more difficult the material, the more pre-reading will help your comprehension. Additionally, pre-reading provides an advance sampling of the information in the material. As we indicated earlier, there is an overload of material of Fire Service information to digest. Pre-reading will give you an idea of whether or not you want to spend any time on the material.

Another advantage of pre-reading is that it makes clear the author's overall point of view and conclusions. This knowledge will help you read the work more critically when you re-read the material more thoroughly. To repeat the concept of pre-reading, the first and third steps in pre-reading an article or the chapter of a book, takes advantage of a nearly universal format of organization of the material. It must be remembered that these divisions or organizational format is not always easily identified, but they exist in all competent writing.

The middle step in pre-reading a chapter or article, is to read the first sentence of each paragraph between the introduction and the conclusion. In the middle paragraphs the author develops what he has to say. The reason that you read the first sentence in every paragraph is because the first sentence is the "topic sentence". The topic sentence contains the paragraph principal thought. By reading the first sentence of each paragraph between the introduction and conclusion, you will obtain a clear, though highly simplified, notion of what the authors principal thoughts are.

SUMMARY

One of the keys to improve examination scores is to increase reading skills. Pre-reading is the technique of looking over the material before reading it thoroughly, making a map to guide you through the unknown territory ahead. Pre-reading will allow you to read more, retain more and provide a method to review important points of a chapter quickly just prior to the examination.

Remember practice makes perfect. Practice is best done for a short time every day rather than long periods once or twice a week.

READING COMPREHENSION DIAGNOSTIC TEST

INSTRUCTIONS:

You have 25 minutes to complete the following reading comprehension diagnostic test.

Select the answer which best answers the question. Base your selection solely on the material in the statements, and not on prior knowledge.

STATEMENT #1

Soviet Asia, the vast land mass between the Urals and the Pacific, between the Arctic Ocean and the borders of China and Afghanistan, is a vividly interesting area. It contains an enormous variety of regions and cultures. The frozen tundra toward the Arctic, for instance, the pioneer Russian farm settlements, the old colorful oases of Inner Asia, with such romantic cities as Smarkland and Bokhara. There is a mosaic of peoples and tribes of the Russian Revolution and the tendencies toward modernization and industrialization which it has initiated.

QUESTIONS 1,2, and 3 ARE BASED ON STATEMENT #1

1. The title below that best expresses the ideas of statement #1 is:

    A. West of the Urals.
    B. Russian Pioneers.
    C. Land of Romance.
    D. Asiatic Russia.

2. The aspect of Soviet Asia with which this passage is primarily concerned is its:

    A. Vast area.
    B. Variety.
    C. Modernization.
    D. Romance.

3. One element generally true of existence in Soviet Asia is:

    A. A dry climate.
    B. Cold weather
    C. A trend toward westernization.
    D. The many pioneer farms.

STATEMENT #2

A hundred years ago, when there were only half as many inhabitants in Europe, the best minds shared the gloomy view of Thomas Robert Malthus. He believed that human population always increases faster than the food supply and that misery and want and war are the inevitable consequences. Three hundred years ago, when the population was about the same as it had been for thousands of years, famine was a periodic experience which came so regularly that man accepted it as normal, like the succession of the seasons. "Seven famines and ten years of famine in a century" was the "law regulating scarcities" prior to 1600, and men accepted it as some now accept the "law of business cycles."

QUESTIONS 4,5, and 6 ARE BASED ON STATEMENT #2

4. The title below that best expresses the ideas of statement #2 is:

   A. The growth of population.
   B. The cause of famine.
   C. The recurrence of famine.
   D. Depression cycles.

5. Malthus' theory was accepted because:

   A. People were more ignorant in those days.
   B. Famine had been life in Europe.
   C. It was an excuse for wars.
   D. It made some people rich.

6. Malthus' theory was:

   A. Inspiring.
   B. Fascinating.
   C. Contradictory.
   D. Discouraging.

STATEMENT #3

Geometry is a very old science. We are told by Herodotuc, a Greek historian, that geometry has its origin in Egypt along the banks of the river Nile. The first record we have of its study is found in a manuscript written by Ahmes, an Egyptian scholar, about 1550 BC. This manuscript is believed to be a copy of a treatise which dated back probably more than a thousand years, and describes the use of geometry at that time in a very crude form of surveying or measurement. In fact, geometry, which means "earth measurement" received its name in this manner. This re-measuring of the land was necessary due to the annual overflow of the river Nile and the consequent destroying of the boundaries of farm lands. This early geometry was very largely a list of rules or formulas for finding the areas of plane figures. Many of these rules were inaccurate, but, in the main, they were fairly satisfactory.

QUESTIONS 7, 8 and 9 ARE BASED ON STATEMENT #3

7. The title that best expresses the ideas of statement #3 is:

    A. Floods of the River Nile.
    B. Beginnings of Geometry.
    C. Manuscript of Ahmes.
    D. Surveying in Egypt.
    E. Importance of the Study of Geometry.

8. In developing geometry, the early Egyptians were primarily concerned with:

    A. Discovering why formulas used in measuring were true.
    B. Determining property boundaries.
    C. Measuring the overflow of the Nile.
    D. Generalizing formulas.
    E. Constructing a logical system of Geometry.

9. One of the most important factors in the development of geometry as a science was:

    A. The inaccuracy of the early rules and formulas.
    B. Annual flooding of the Nile Valley.
    C. Destruction of farm crops by the Nile.
    D. An ancient manuscript copies by Ahmes.

STATEMENT #4

The characteristic American believes, first, in justice as the foundation of civilized government and society, and, next, in freedom for the individual, so far as that freedom is possible without interference with the equal rights of others. He conceives that both justice and freedom are to be secured through popular respect for laws enacted by the elected representatives of the people and through the faithful observance of those laws. It should be observed, however, that American justice in general keeps in view the present common good of the vast majority, and the restoration rather than the punishment of the exceptional malignant or defective individual. It is essentially democratic; and especially it finds sufferings inflicted on the innocent unintelligible and abhorrent.

Blind obedience and implicit submission to the will of another do not commend themselves to characteristic Americans. The discipline in which they believe is the voluntary cooperation of many persons in the orderly and effective pursuit of common ends. Thus they submit willingly to any restrictions on individual liberty which can be shown to be necessary to the preservation of the public health, and they are capable of the most effective cooperation needed in business, sports, and war.

QUESTIONS 10, 11, 12 and, 13 ARE BASED ON STATEMENT #4

10. The title below that best expresses the ideas of statement #4 is:

    A. American Justice.
    B. A plea for Cooperation.
    C. The basis of American Democracy.
    D. The American Government.

11. The primary element in the American way of life is:

    A. The right to vote.
    B. Freedom.
    C. Unwillingness to follow leaders.
    D. Justice.
    E. Popular respect for laws.

12. American justice emphasizes:

    A. The welfare of the minority.
    B. Retaliation for disobedience.
    C. Rehabilitation of wrongdoers.
    D. The sufferings of the innocent.
    E. Punishment of criminals.

13. The American people believe in:

    A. Strict discipline.
    B. Liberty without restraint.
    C. Subservience to the President.
    D. Working together for a necessary purpose.

STATEMENT #5

In the South American rain forest abide the greatest acrobats on earth. The monkeys of the Old World, agile as they are, cannot hand by their tails. It is only the monkeys of America that possess this skill. They are called ceboids, and their unique group includes marmosets, owl monkeys, sakis, spider monkeys, squirrel monkeys, and howlers. Among these, the star gymnast is the skinny, intelligent spider monkey. Hanging head down like a trapeze artist from the loop of a liana, he may suddenly give a short swing, launch himself into space and, soaring outward and downward across a 50-foot void of air, lightly catch a bough on which he spied a shinning berry. No owl monkey can match this leap, for their arms are shorter, their tails untalented. The marmosets, smallest of the tribe, tough noisy hoodlums that travel in gangs, are also capable of leaps into space, but their landings are rough; smack against a tree trunk with arms and legs spread wide.

QUESTIONS 14, 15, and, 16 ARE BASED ON STATEMENT #5

14. The title below that best expresses the ideas of statement #5 is:

    A. The Star Gymnast.
    B. Monkeys and Trees.
    C. The Uniqueness of Monkeys.
    D. Ceboid Acrobats.

15. The author states that the marmoset is:

    A. Clannish.
    B. Graceful.
    C. Antisocial.
    D. Shy.
    E. Intelligent.

16. Compared with monkeys of the Old World, American monkeys are:

    A. Smaller.
    B. More quite.
    C. More dexterous.
    D. More protective of their young.
    E. Less at home in their surroundings.

ANSWERS

READING COMPREHENSION DIAGNOSTIC TEST

1. D
2. B
3. C
4. C
5. B
6. D
7. B
8. C
9. A
10. A
11. D
12. C
13. D
14. D
15. A
16. C

CHAPTER 8

READING COMPREHENSION: CONCENTRATION AND MEMORY

READING COMPREHENSION: CONCENTRATION AND MEMORY

INTRODUCTION

The ability to concentrate and memorize are essential keys for reading comprehension. Like anything else in life, concentration and memory when reading can be analyzed, practiced and improved. Many people have difficulty with concentration when noise, other outside distractions,,or something else on your mind seems to make it impossible to keep your attention on the page for more than a paragraph or two at a time. In some cases you might have read a chapter in a book and when you finished you asked yourself: "what have I read". The obvious problem in this case is the inability to concentrate. This is one of the barriers to speed, concentration, and retention.

This session will discuss steps that can be taken to improve concentration and memory.

READING CONCENTRATION:

Because of the tremendous volume of Fire Service material available, the ability to concentrate on the material is essential to achieve your goal to become a Firefighter. Yet poor concentration is one of the most common problems in reading. When concentration is limited, the difficulty in remembering is increased. Some of the environmental aspects that affect the ability to concentrate are noise, personal matters on your mind, boring material, difficult material, poorly written material or you are just plain tired.

Students have complained many times that they have had occasions where they have read an entire chapter in a book and have little idea as to what they have read. Whenever this happens, the inability to concentrate is usually the barrier to speed, comprehension, and retention.

For most people, the ability to concentrate depends on the balance of interest at the moment. This means that there is a competition between your reading material and other things you might want to give attention. When this happens and you are leaning away from your reading material, remember this: The reason you are doing this reading is to learn enough to get a job as a Firefighter. This thought should always be in your mind when you are identifying the purpose of the reading and you are starting to think of other non Fire Service related things to do.

Concentration improves when you are in a good reading environment. The obvious question that follows is, what is a good reading environment? A good reading environment is a quiet place, free from distraction and free from interruption by other people. This environment can be found in a library; but less formal, less public, more accessible, and emotionally warmer places may prove to be more conducive to reading concentration. Your reading should be done in a comfortable but firm, upright chair. It has been said that you learn better while experiencing some degree of bodily tension. You know what can happen in a chair that is too comfortable. When reading a book place it on a desk or table in front of you. Support the book with a slanted bookrest, or drop it against another book, so that the top of the page is no farther away from your eyes than the bottom. This prevents your eyes from having to adjust from far to near as you read down the page.

Some readers who do most of their reading sitting down will occasionally stand to read for a short time. This is one way of combatting fatigue or sleepiness.

Lighting is adequate if you can read without strain. Not only the book, but the rest of the room as well, should be lit. If the lighting is properly balanced, there will be no glare on the page, and no sense of marked contrast between the lighted area and the rest of the room.

Reading for several hours without a break can cause eye fatigue and discomfort. Therefore, remember to rest your eyes about once each half hour by closing them or looking at a distant object for a few minutes. You will then return to your reading refreshed, and your reading efficiency will rise.

It has been indicated that eating candy or some other high sugar content food helps restore energy. Recent test have indicated this is not the best solution to restore energy. The reason is that the sugar in the candy only gives you a very temporary energy increase. The energy increase is followed by a rapid energy decrease below the level that you started. Studies have indicated a brisk 5 minute walk is a much better means of restoring energy.

Another question often asked is does reading after a meal effect reading comprehension? The answer is yes. Trying to read after a meal is difficult because all the energy in your body is dedicated to digest the food in your stomach. The best procedure is to eat a light lunch or dinner when you know you have a lot of reading to do afterwards.

READING MEMORY:

A good technique to use when trying to remember names or particular facts is the linkage by association of a new name or a fact to a known name or fact.

Repetition helps increase memory. The more exposure you have to the material, the more you will remember. There is a concept called short term memory and long term memory. Short term memory, for example, is when you look up a telephone number in the directory, then hold it in mind while you dial it. By the time you have hung up at the end of the conversation you have forgotten the number. Long term memory is retaining information on a more permanent basis. The key for you, as an entrance Firefighter candidate, is to transfer the Fire Service information you are learning from short term memory to long term memory.

Keys to help you transfer information from your short term memory to your long term memory are utilizing all the methods previously discussed. Association of an unknown to a known. Concentrating on the material. Concentration is increased by your reading environment, frame of mind, and by reading faster. The more times you are exposed to the material or information, the more you will retain. Remember pre-reading is a good source for a review of a chapter or book by providing exposure to the materials key points and major theme.

SUMMARY

In this session we have discussed various concepts to help increase reading comprehension. We have keyed on concentration and memory. There can be two types of distraction that upset your balance of interest and attention. The first is external: your reading environment and the temptations it offers to lure your mind away from the reading material. The external extractions can usually be controlled by a few simple precautions.

The second type is "what's on your mind". The competing thoughts that preempt the attention you should be giving to the material at hand. The latter is perhaps the most difficult to deal with of all the causes of poor concentration. Many of the techniques you have been learning can be used to increase your interest in the material, and thus help overcome distraction you cannot otherwise eliminate.

ASSIGNMENT:

1. Read "Fire Pumps": Information Sheet #10. Pages 358 - 363.
2. Prepare for a quiz on this material.

READING COMPREHENSION QUESTIONS:

STATEMENT

Specific gravity equals the weight or mass of a given volume of a substance at a specified temperature as compared to that of an equal volume of another substance. Specific gravity applies to liquids only. Liquids of a specific gravity of less than 1 will float on water. The specific gravity of water is equal to 1 at 4 degrees Centigrade or 39 degrees Fahrenheit. Specific gravity is a ratio of weight to volume. When calculating the specific gravity of a substance, you are determining the ratio of the weight of a solid or liquid substance to the weight of an equal volume of water. The ratios are compared to water because it is simple to determine if the substance sinks or floats.

ACCORDING TO THE PREVIOUS STATEMENT

1. When comparing specific gravity:
    A. It is compared to mercury.
    B. It is compared to water.
    C. It is compared to any substance of unequal ratios.
    D. All of the above.

2. The specific gravity of water is equal to:
    A. 1.
    B. 5.
    C. 10.
    D. 15.

3. Specific gravity is a ratio of:
    A. Length to height.
    B. Weight to length.
    C. Weight to volume.
    D. Height to volume.

4. The term "specific gravity" applies to:
    A. Gases.
    B. Liquids.
    C. Solids.
    D. All of the above.

5. Liquids of a specific gravity of:
    A. Less than 1 will float on water.
    B. More than 1 will float on water.
    C. Less than 1 will sink in water.
    D. All of the above.

ANSWERS:

1. B 2. A 3. C 4. B 5. A

STATEMENT

Vapor density equals the weight of a vapor-air mixture resulting from the vaporization of a flammable liquid at equilibrium temperature and pressure conditions, as compared with the weight of an equal volume of air under the same conditions. Vapor density is the relative density of vapor gas, with no air present, compared to air. Vapor density is a ratio of gases. The vapor density of air is equal to 1, gases of less than 1 indicates that the gas is lighter than air. In order to compare vapor densities of gases, you must know that the molecular weight of air is equal to 29.

ACCORDING TO THE PREVIOUS STATEMENT

1. Vapor density of air is equal to:
    A. 15.
    B. 10.
    C. 5.
    D. 1.

2. The molecular weight of air is equal to:
    A. 1.
    B. 11.
    C. 19.
    D. 29.

3. Vapor density is equal to a ratio of:
    A. Gases.
    B. Liquids.
    C. Solids.
    D. All of the above.

4. Flammable liquid vapors have a vapor density:
    A. Of less than 1 and will sink in the atmosphere.
    B. Of greater than 1 and will sink in the atmosphere.
    C. Of greater than 1 and will rise in the atmosphere.
    D. None of the above.

5. Vapor density ratios are compared:
   A. To air.
   B. To solids.
   C. To liquids.
   D. In unlike conditions.

ANSWERS:

1. D   2. D   3. A   4. B   5. A

STATEMENT

There are two modes of which fire burns, the flaming mode and the smoldering mode. There are three stages of fire, the first stage of fire is a smoldering or incipient phase where the oxygen level is at 21% and the fire is burning at 100 degrees F., with a room temperature of 100 degrees F. The second stage is the flame producing phase where the oxygen level is at 21% to 15% with a fire and room temperature of 1300 degrees F. The third stage is a smoldering phase, or a decrease in heat generation, where the oxygen level is below 15% and the fire and room temperature is at 1000 degrees F.

ACCORDING TO THE PREVIOUS STATEMENT

1. How many modes are there that fire will burn?
    A. There are two modes at which fire burns.
    B. There are three modes at which fire burns.
    C. There are four modes at which fire burns.
    D. There are five modes at which fire burns.

2. How many stages of fire are there?
    A. There are three stages of fire.
    B. There are four stages of fire.
    C. There are five stages of fire.
    D. There are six stages of fire.

3. During the:
    A. First stage of fire the room temperature = 1000 degrees F.
    B. First stage of fire the room temperature = 100 degrees F.
    C. Second stage of fire the room temperature = 100 degrees F.
    D. Third stage of fire the room temperature = 100 degrees F.

4. During the:
    A. First stage of fire the temperature of the fire = 100 degrees F.
    B. Second stage of fire the temperature of the fire = 1000 degrees F.
    C. Third phase of fire the temperature of the fire = 100 degrees F.
    D. Third phase of fire the temperature of the fire = 1000 degrees F.

5. During the:
    A. First phase of fire the oxygen content = 15%.
    B. First phase of fire the oxygen content = below 15%.
    C. Second phase of fire the oxygen content = 21% to 15%.
    D. Third phase of fire the oxygen content = 21%.

ANSWERS:

1. A   2. A   3. B   4. D   5. C

STATEMENT

During the first stage of fire there is little or no decrease in the oxygen content of the interior atmosphere or in the average temperature of the interior atmosphere. Major damage will be caused by smoke during the first stage of fire. During the second stage of fire is when the major destruction will take place. During the third stage of fire there is a possibility of an inward rupture of window panes. If a Firefighters opens a door to an apartment and a smoke explosion occurs, the fire was most likely in the third stage of fire.

ACCORDING TO THE PREVIOUS STATEMENT

1. During the:
    A. First stage of fire there is a large increase in the oxygen content.
    B. First stage of fire there is a large increase in temperature.
    C. First stage of fire there is no decrease in oxygen content.
    D. None of the above.

2. Major destruction:
    A. Takes place during the first stage of fire.
    B. Takes place during the second stage of fire.
    C. Takes place during the third stage of fire.
    D. Rarely takes place during any stage of a fire.

3. During the:
    A. First stage of fire there is the possibility of the inward rupture of windows.
    B. Second stage of fire there is the possibility of the inward rupture of windows.
    C. Third stage of fire there is the possibility of the inward rupture of windows.
    D. All of the above.

4. During the:
    A. First stage of fire there will be major damage by smoke.
    B. First stage of fire there will be no damage by smoke.
    C. Any stage of fire there will be no smoke damage created.
    D. Both B and C.

5. If a firefighter opens the door to an apartment and an explosion occurs, the fire was most likely in the:
    A. First stage of fire.
    B. Second stage of fire.
    C. Third stage of fire.
    D. Between the first and second stage of fire.

ANSWERS:

1. C   2. B   3. C   4. A   5. C

STATEMENT

Many people are not aware of the duties Firefighters perform, such as: training, fire prevention, and hydrant maintenance, etc. The general public usually are of the opinion that Firefighters are just sitting around the firehouse between fires. There are ways that Firefighters can assist in changing this image, such as: appearance, by greeting visitors who come to the firehouse, by their behavior on the street at a fire, and by treating the public in a courteous manner. Example:
75% of the rescues made by the average Fire Department take place at relatively small fires, not at dramatic large alarm fires. The public seldom hears about them because the Fire Departments seldom let the media know about Firefighters who have executed acts of heroism at routine fires.

ACCORDING TO THE PREVIOUS STATEMENT

1. When not fighting fires, Firefighters:
    A. Lounge around the firehouse.
    B. Work on public relation projects.
    C. Repair tools and equipment.
    D. None of the above.

2. The responsibility of improving the Fire Departments image should be placed on:
    A. The media.
    B. The public.
    C. The people rescued by Firefighters.
    D. The Fire Department.

3. Most rescues made by the Fire Department take place at:
    A. Large alarm fires.
    B. Special emergencies where no fire is involved.
    C. Relative small fires.
    D. Dramatic large fires.

4. The public rarely hears about rescues made by Firefighters at routine fires because:
    A. Information about fires must be kept confidential.
    B. Fire Departments seldom report these rescues to the media.
    C. Most of these rescues take place late at night.
    D. Reporters are not interested in covering routine fires.

5. What would be the best title for this material?
    A. An Inside Look at the Fire Department.
    B. Making the Most of Fire Prevention Week.
    C. Improving the Fire Departments Public Image.
    D. Brave Acts Performed by Firefighters.

ANSWERS:

1. D   2. D   3. C   4. B   5. C

STATEMENT

For many years the fire community acknowledged only three fire classifications. In 1960 the classifications were reorganized to show four fire classifications:
1. Class A fires = ordinary combustibles.
2. Class B fires = flammable liquids/gases.
3. Class C fires = electrical.
4. Class D fires = combustible metals.

CLASS "A" FIRES: these fires include ordinary combustibles such as wood, paper, fabric, solid plastics, and rubber. Class A fires normally involve fuels of an organic nature. These fires are the most common. Extinguishing agents for Class A type fires include water, some foam types, and multipurpose extinguishers.

CLASS "B" FIRES: these fires include all flammable and combustible liquids, gases, greases, and oils. One way to recognize a Class B fuel is by the container. No Class B fuel retains its own shape, because they are liquids and gases, These types of materials are usually stored in a strong rigid container. Extinguishing agents for Class B type fires include carbon dioxide, dry chemical, and foam types.

CLASS "C" FIRES: a Class C fire is one that involves energized electrical equipment. Very special importance must be given to the electrical non-conductivity of the extinguishing agent. Only when the electrical circuits have been de-energized may Class A and Class B extinguishing agents be used. Extinguishing agents suitable for Class C fires include dry chemical, carbon dioxide, and halon types.

CLASS "D" FIRES: when metal burns, they pose some very unique hazards. They burn extremely hot. They may actually react to ordinary extinguishing agents. Class D fires are fires involving such metals as sodium, magnesium, aluminum, uranium, and titanium. The hazards of a metal fire are so unique that ordinary extinguishing agents should generally not be used. Instead, extinguishing agents for Class D fires are those that have been specifically designed and approved for that type of application, such as dry powder types, Metal "X", etc.

ACCORDING TO THE PREVIOUS STATEMENT

1. There are how many classification of fire?
    A. 3.
    B. 4.
    C. 5.
    D. 6.

2. Class A fires are fires involving:
    A. Ordinary combustibles.
    B. Flammable liquids/gases.
    C. Energized electricity.
    D. Combustible metals.

3. Class B fires are fires involving:
    A. Ordinary combustibles.
    B. Flammable liquids/gases.
    C. Energized electricity.
    D. Combustible metals.

4. Class C fires are fires involving:
    A. Ordinary combustibles.
    B. Flammable liquids/gases.
    C. Energized electricity.
    D. Combustible metals.

5. Class D fires are fires involving:
    A. Ordinary combustibles.
    B. Flammable liquids/gases.
    C. Energized electricity.
    D. Combustible metals.

6. Class A fires include combustibles such as all of the following, EXCEPT:
    A. Wood.
    B. Gases.
    C. Paper.
    D. Rubber.

7. Class B fires include combustibles such as all of the following, EXCEPT:
    A. Combustible liquids.
    B. Rubber.
    C. Greases.
    D. Gases.

8. Extinguishing agents for Class A type fires include all of the following, EXCEPT:
    A. Water.
    B. Some foam types.
    C. Dry chemical.
    D. CO2.

9. Extinguishing agents for Class B type fires include all of the following, EXCEPT:
    A. Carbon dioxide.
    B. Halon types.
    C. Dry chemical.
    D. All of the above.

10. Extinguishing agents for Class C type fires include all of the following, EXCEPT:
    A. Dry chemical.
    B. Carbon dioxide.
    C. Halon types.
    D. Dry powder.

11. Class C fires include:
    A. Wood.
    B. Gases.
    C. Energized electricity.
    D. Petroleum products.

12. Class D fires include combustible metals such as:
    A. Sodium.
    B. Magnesium.
    C. Titanium.
    D. All of the above.

ANSWERS:

1. B	4. C	7. B	10. D
2. A	5. D	8. D	11. C
3. B	6. B	9. B	12. D

STATEMENT

Fire Hose made of linen and unlined is basically tightly woven linen thread in the shape of a pipe. Because of the inherent properties of linen, soon after water has passed through a linen fire hose, the threads will expand causing the small spaces between them to close, making the hose watertight. Linen fire hose is inclined to deteriorate quickly if it is not completely dried after use or if located where it will be subjected to dampness or the climate. Linen fire hose is not normally constructed to withstand constant use or to be used where the material will be subjected to chafing from jagged or sharp areas.

ACCORDING TO THE PREVIOUS STATEMENT

1. Soon after water has passed through linen hose:
    A. It will become water tight.
    B. Small spaces will close.
    C. Will continue to leak a large amount of water.
    D. Both answers A and B are correct.

2. Unlined linen fire hose is best suited for use:
    A. In an area where it will not be subjected to sharp objects.
    B. For use at the fire academy.
    C. As standard equipment on fire apparatus.
    D. As a garden hose.

3. The LEAST appropriate use for unlined linen fire hose would be:
    A. Emergency fire hose in a grocery store.
    B. Emergency fire hose in an office building.
    C. Emergency fire hose at a lumber yard.
    D. Emergency fire hose in a warehouse.

4. Unlined linen fire hose is made of:
    A. Plastic.
    B. Rubber.
    C. Synthetics.
    D. Fabric.

5. Unlined linen fire hose is best if used:
    A. Frequently.
    B. Infrequently.
    C. in rough terrain.
    D. Daily.

ANSWERS:

1. D   2. A   3. C   4. D   5. B

STATEMENT

Fire hose has consistently been the most used piece of equipment in the Fire Service. Water is the extinguishing agent most frequently used in fighting fires. Fire hose is the means of moving water from the source to the area of need. Fire hose is commonly accessible in the following sizes and is measured by internal diameter:

1. 3/4 inch.
2. 1 inch.
3. 1 1/2 inch.
4. 2 1/2 inch.
5. 3 1/2 inch.
6. 4 inch.
7. 5 inch.
8. 6 inch.

Fire hose is produced in many types of construction, based upon the expected field of service. The most conventional type of construction incorporates a rubber liner and one or two woven or molded outside covers. Some types of fire hose may be used for many different situations. Usage usually is in the category of suction or discharge.

ACCORDING TO THE PREVIOUS STATEMENT

1. The most used piece of equipment in the Fire Service is:
   A. Water.
   B. Ladders.
   C. Fire hose.
   D. Fire extinguisher.

2. The most common extinguishing agent used in fighting fires is:
   A. Water.
   B. Foam.
   C. Dry chemical.
   D. Fire hose.

3. Fire hose is generally available in all of the following sizes, EXCEPT:
   A. 4 inch.
   B. 5 inch.
   C. 6 inch.
   D. 8 inch.

4. The most common type of fire hose construction is:
   A. Dacron.
   B. Unlined.
   C. Rubber lined with two woven covers.
   D. Rubber lined with one woven cover.

5. Usage of fire hose can be divided into two types, which are:
   A. Pressure and non-pressure.
   B. Suction and discharge.
   C. Suction and non-pressure.
   D. Discharge and pressure.

ANSWERS:

1. C   2. A   3. D   4. C   5. B

STATEMENT

Fire hose is designed for the use of transporting water and should never be utilized for anything other than that function. The demand for routine testing and maintenance is evident. A burst or ruptured hose at a fire scene may be life threatening. Not only does this stop the water supply, but it creates a significant hazard to those in the area because of the flailing action of the hose that is out of control. In order to help insure that fire hose will meet performance standards, fire hose should be tested at least annually.

The most commonly used fire hose is the woven-jacket type, with the possible exception of booster and hard suction hoses. Fire hose should be loaded on apparatus so as to allow air to circulate freely, although this is not always feasible. High humidity climates make this a more complex problem to deal with. Fire hose should be removed from apparatus and reloaded, at certain time intervals, in a different order so that the hose that was on the bottom will be in a different position and all the hose should be in a different order, so that the kinks may be eliminated. When hose is subjected to rain or snow, it should be placed on wooden racks with an air space below. The hose bed and all of the exposed fire hose should be covered with some sort of a waterproof cover. A waterproof cover not only protects the fire hose from moisture, they also will reduce deterioration of the fire hose from continued exposure to the sun. Fire hose should be thoroughly cleaned and dried periodically.

During emergency incidents, fire hose should be positioned without sharp kinks that can cause extreme internal stress. The proper positioning of fire hose should include avoiding sharp or rough objects. Except in an emergency situation, vehicles should never drive directly over fire hose. A vehicle may drive over fire hose with the use of a hose bridge, which will keep the weight of the vehicle off of the fire hose, thus the vehicle will not cause any damage to the fire hose. Also hose belts or rope tools should be used whenever fire hose is hoisted up ladders or other elevations, so as to take the excessive stress off of the couplings.

Firefighters must be aware of the damage that water hammer can cause to fire hose. Water hammer can take place when a valve or nozzle is closed too quickly. The water will continue to move within the fire hose which will cause increased pressure. This increased pressure could damage the fire pump or cause the fire hose to rupture. Therefore, firefighters should always close valves and nozzles slowly.

STATEMENT
(continued)

To protect the discharge side of the pump, pressure surges are averted by the use of relief valves and speed governors in union with pumping operations. During relay or tandem pumping this is particularly important. Relay and tandem pumping is needed when supplying water through fire hose in lengths of over 2000 feet or when delivering water up a grade. During this procedure an pumper is placed at the water source and delivers water to another pumper or to the fire, depending on the length of the hose lay. Relay or tandem pumping is needed because of the added friction loss created by the fire hose over a long distance.

Special care should be taken, during cold climate conditions, to prevent water from freezing inside the fire hose. After the flow of water is initiated, a constant flow of water should continue until the fire hose is not needed and is going to be removed. Avoid sharp bends when ice has developed. Ice inside fire hose may act like a knife. It may be necessary to chop frozen fire hose out of the ice. Care must used when doing this so as not to damage hose.

Firefighters should avoid placing fire hose too close to the fire, because the fire hose may be scorched or burned. In wildland fire conditions, the use of unlined fire hose is a benefit because water will ooze through the fabric and obstruct the material from burning by keeping the fabric wet.

After fire incidents, fire hose should be laid out, at the fire station, and brushed cleaned with a mild soap and water to remove dirt, etc. The use of a small hose will be useful during the cleaning procedure. The fire hose must be thoroughly dried after it is cleaned. Fire hose may be dried by hanging it in a hose tower, air drying on horizontal racks, or in a heated forced air drying machine. Dry fire hose should be rolled and placed on racks. Rubber fire hose, such as booster hose only needs to have dirt, etc. wiped off. Many of the larger Fire Departments have a central fire hose cleaning and maintenance facility.

As far as fire hose is concerned, good record keeping is vital along with the normal use, maintenance, and storage. Each length of fire hose should be identified with a number. When hose changes take place, the lengths and order of fire hose on equipment should be recorded. This will allow firefighters to calculate the length of hose leads by looking at the inventory number on the fire hose and comparing it with the inventory list of the apparatus. Fire hose certification test should also be recorded and periodic replacement may be sanctioned.

ACCORDING TO THE PREVIOUS STATEMENT

1. Fire hose is designed for the use of:
    A. Adding pressure to the water.
    B. Reducing water pressure.
    C. Moving water.
    D. Holding water.

2. Fire hose should be tested at least:
    A. Twice a year.
    B. Once a year.
    C. Once every two years.
    D. Does not need to be tested.

3. It is best to load fire hose:
    A. So that air will circulate.
    B. Tightly.
    C. Always in the same position.
    D. Always in the same order.

4. Fire hose should be:
    A. Covered in the hose bed.
    B. Cleaned and thoroughly dried.
    C. Protected from the sun.
    D. All of the above.

5. As far as vehicles driving over fire hose, which of the following is true:
    A. Should never happen.
    B. Will not cause any damage.
    C. Is alright only during drills.
    D. Should never happen, except in an emergency.

6. The condition that occurs when a valve or nozzle is abruptly closed, while water continues to move in the hose, causing increased pressure and may damage or rupture the hose is called:
    A. Head pressure.
    B. Back pressure.
    C. Water Hammer.
    D. Water damage.

7. To prevent pressure surges during pumping operations, relief valves and engine speed governors are operated:
    A. In union with pumping operations.
    B. Opposite of each other.
    C. At all times.
    D. None of the above.

8. Tandem or relay pumping is necessary when supplying water through hose lengths of over:
   A. 1000 feet.
   B. 1200 feet.
   C. 1500 feet.
   D. 2000 feet.

9. Relay pumping is basically necessary because:
   A. Of Head pressure.
   B. Of Back pressure.
   C. Of Friction.
   D. Nozzle reaction.

10. To prevent water from freezing inside fire hose in cold climate conditions: (water flowing)
    A. Shut off hose line when it is not being used.
    B. Continue flowing water even when hose line is not being used.
    C. Use increased pressures.
    D. Used decreased pressures.

11. In wildland fires, what helps prevent unlined fire hose from being burned?
    A. Water leaking through the fabric.
    B. The added insulation.
    C. Water accumulation in the area.
    D. None of the above.

12. After fire incidents, fire hose should be:
    A. Rinsed with clear water.
    B. Rinsed with mild soap.
    C. Brushed and dried.
    D. All of the above.

13. Rubber hose, such as booster hose, should be:
    A. Washed and dried.
    B. Dried only.
    C. wiped clean.
    D. None of the above.

14. In addition to normal usage, cleaning, and storage of fire hose, it is necessary to keep:
    A. Records of fire hose.
    B. Old fire hose.
    C. Old couplings.
    D. None of the above.

ANSWERS:

1.C	4. D	7. A	10. B	13. C
2. B	5. D	8. D	11. A	14. A
3. A	6. C	9. C	12. D	

STATEMENT

In the fire service, ventilation is the method of opening up a building or structure that is involved in fire so as to liberate the accumulated smoke, gases, and heat. If Firefighters do not have the required knowledge of the principles of ventilation, it could create unnecessary damage, and or injuries. Ventilation by itself will not extinguish fires, but when used in an prudent manner it will allow Firefighters to locate the fire easier and faster with less difficulty and risk.

ACCORDING TO THE PREVIOUS STATEMENT

1. The major result of failing to apply the principles of ventilation at a fire may be:
    A. Injury to the Firefighters.
    B. Inappropriate use of equipment and apparatus.
    C. Waste of water.
    D. Disciplinary action.

2. The best reason for ventilation is that it:
    A. Will lower the temperature of the fire.
    B. Eliminates the need for breathing apparatus.
    C. Allows the Firefighters to advance closer to the fire.
    D. Reduces smoke damage.

3. The use of ventilation principles would be the LEAST useful in a fire involving:
    A. Apartment building.
    B. Grocery market.
    C. Bank.
    D. Lumber yard.

4. For a well trained and well equipped Firefighter, ventilation is:
    A. Not needed.
    B. Unimportant.
    C. A basic procedure.
    D. Rarely used.

5. Ventilation of a fire building or structure will release:
    A. Excessive water.
    B. Occupants.
    C. Smoke and heat, but not the toxic gasses.
    D. Smoke, heat, and the toxic gases.

ANSWERS:

1. A   2. C   3. D   4. C   5. D

STATEMENT

Within the Fire Service, the Hurst tool and shears have many uses in both rescue and forcible entry. Each tool, the "jaws of life" and the shears, may be used individually or simultaneously from the same power supply.

The power supply produces hydraulic oil pressure that circulates through flexible lines. These lines must be attached to the tools before they will function. The fluid used for the tools is not a standard hydraulic oil, it is an oil of an acid base that can cause burns to the skin, if contacted.

Each tool weighs about 60 pounds. The power unit weighs about 40 pounds. The flexible arms of the tool are about 2 1/2 feet long and made of forged titanium. The tensile strength of the arms is 155,000 pounds. The actual force exerted at the tip of the "jaws" is 10,000 pounds in both the push or pull operation.

The "jaws of life" are used for prying and separating, without the ability to perform cutting operations.

To operate the tool firefighters must grasp the back portion with both hands, using the thumb on the right hand to control the opening and closing of the jaws. The control must also be operated by left handed firefighters with their right thumb. Firefighters may cut the post, which anchors a car top to the body of the car, with one cutting motion, with the use of the "shears". When this is completed firefighters may extricate victims from the car by literally peeling the car top back.

Firefighters should always be aware of sharp and jagged edges while operating the shears. Using these tools, firefighters can accomplish, in seconds, cutting operations which include windshield post and metal-clad door frames.

Firefighters, with the added use of chains, can pull a steering wheel and column completely out of a car. The tool is capable of lifting the weight of a car or a truck.

ACCORDING TO THE PREVIOUS STATEMENT

1. The hurst tool and shears:
    A. Must be operated separately.
    B. Must be operated together.
    C. May be operated separately or together from one power supply.
    D. If operated together, must have two power supplies.

2. The power supply for the hurst tool will generate:
    A. Vacuum pressure.
    B. Air pressure
    C. Hydraulic oil pressure.
    D. Water pressure.

3. The hurst tool weighs about:
    A. 40 pounds.
    B. 50 pounds.
    C. 60 pounds
    D. 70 pounds.

4. The jaws of the hurst tool are flexible and made of forged:
    A. Aluminum.
    B. Steel.
    C. Magnesium.
    D. Titanium.

5. The length of the jaws on the hurst tool are about:
    A. 1 1/2 feet long.
    B. 2 feet long.
    C. 2 1/2 feet long.
    D. 36 inches long.

6. The tensile strength of the jaws are:
    A. 55,000 pounds.
    B. 75,000 pounds.
    C. 100,000 pounds.
    D. 155,000 pounds.

7. The actual force exerted at the tip of the jaws is:
    A. 5,000 pounds push and 10,000 pounds pull.
    B. 5,000 pounds pull and 10,000 pounds push.
    C. 10,000 pounds push and 10,000 pounds pull.
    D. 5,000 pounds push and 5,000 pounds pull.

8. To control the opening and closing of the jaws, the firefighter must use:
    A. Right hand thumb.
    B. Left hand thumb.
    C. Right hand or left hand thumb.
    D. Left hand thumb, if the firefighter is left handed.

9. When cutting operations are necessary:
    A. Increase the power of the tool.
    B. Decrease the power of the tool.
    C. Use the shears.
    D. Both A and B.

10. With the use of the hurst power and chains, it is possible to:
    A. Pull a steering wheel completely off.
    B. Pull a steering column completely off.
    C. Lift the weight of a car or a truck.
    D. All of the above.

ANSWERS:

| 1. C | 3. C | 5. C | 7. C | 9. C |
| 2. C | 4. D | 6. D | 8. A | 10. D |

STATEMENT

Spontaneous ignition is the ignition due to chemical reaction or bacterial action in which there is a slow oxidation of organic compounds until the material ignites. Usually there is sufficient air for oxidation but not enough ventilation to carry heat away as it is generated. Substances that are susceptible to spontaneous ignition are able to catch fire without an external source of heat. In all cases of spontaneous ignition, heat of oxidation must be produced more rapidly than it is dispersed.

ACCORDING TO THE PREVIOUS STATEMENT

1. Spontaneous ignition is due to:
    A. Chemical reaction.
    B. Bacterial action.
    C. Both A and B.
    D. None of the above.

2. Spontaneous ignition requires that oxidation:
    A. Take place without air.
    B. Take place with enough air to carry heat away as it is generated.
    C. Take place with enough air for the oxidation, but not enough to carry the heat away as it is generated.
    D. None of the above.

3. In all cases of spontaneous ignition the heat of oxidation must be:
    A. Dispersed more rapidly than it is produced.
    B. Produced more rapidly than it is dispersed.
    C. Not produced at all.
    D. Not dispersed at all.

4. Substances that are susceptible to spontaneous ignition are able to catch fire:
    A. Without an external source of heat.
    B. Only with an external source of heat.
    C. Under any conditions.
    D. None of the above.

ANSWERS:

1. C   2. C   3. B   4. A

STATEMENT

Size-up is the mental evaluation made by the fire officer in charge., which enables him to determine a course of action. Size-up is a form of reconnaissance. Size-up includes such factors as time, exposure, property involved, nature and extent of the fire, available water supply and other fire fighting facilities. The size-up report is usually given via radio. In order for a fire officer to make a correct size-up, he should maintain a disciplined mind, and be instructed to think logically during confusion and excitement.

The four stages of size-up include: anticipating the situation, gathering the facts, evaluating the facts, and determining the procedures. Some things to consider at the time of size-up include: the location of the fire, location of the fire within the structure, size of fire, smoke and gases that are being generated, type of contents within the structure, potential of danger to Firefighters, and potential of danger to occupants.

ACCORDING TO THE PREVIOUS STATEMENT

1. Size-up enables the fire officer to determine:
    A. What caused the fire.
    B. Where the fire started.
    C. Course of action.
    D. What is burning.

2. When a fire officer is making a size-up he should consider:
    A. Time of day.
    B. Exposures
    C. Property involved.
    D. All of the above.

3. Size-up is usually given via:
    A. Telephone.
    B. Radio.
    C. Face to face.
    D. Headquarters.

4. All but one of the following are considered stages of size-up, which one is not:
    A. Anticipating the situation.
    B. Gathering the facts.
    C. Evaluating the facts.
    D. Performing the proper procedures.

5. All but one of the following should be considered during size-up, which one is not:
    A. Location of fire.
    B. Size of fire.
    C. Who started the fire.
    D. Smoke and gases being generated.

ANSWERS:

1. C   2. D   3. B   4. D   5. C

STATEMENT

Foam fire extinguisher use three chemicals, bicarbonate of soda, aluminum sulphate and a stabilizer. The small interior section will usually comprise a 50% water solution of aluminum sulphate. The larger outer section will comprise of approximately 7% bicarbonate of soda, 3% stabilizer, and 90% water. The stabilizer is used to make the bubbles smaller in diameter and more persistent.

When the fire extinguisher is inverted the chemicals of the two sections will intermix, which creates the foam and causes it to expel.

The main extinguishing agent is the small bubbles of carbon dioxide gas that will be trapped within the walls of the aluminum hydrate, which forms a hardy, durable, flexible and sticky foam that is able to withstanding a considerable amount of abuse. 90% of the foam consist of carbon dioxide gas volume, although approximately 75% of the foam is water by weight.

The foam will create a blanket of bubbles over the ignited substance, which will eliminate the air and cool the surface. Cellulose products such as fabric are made fire resistant by the strengthened bubbles containing aluminum hydrate. Both horizontal and vertical surfaces are coated and isolated from the heat by the foam adhering wherever it is utilized including floating over liquids.

The foam is harmless to Firefighters. The foam creates less of a wetting effect than water. The extinguisher is not effective on fires involving alcohol. The extinguisher should be protected from low temperatures.

ACCORDING TO THE PREVIOUS STATEMENT

1. The stabilizer in a foam fire extinguisher, has the primary function of preventing:
   A. Clinging of the foam to the extinguisher.
   B. Early chemical reaction of the ingredients.
   C. Swift breaking-up of the carbon dioxide bubbles.
   D. Vaporization of the aluminum sulphate.

2. The primary justification for removing the air from an ignited substance with the use of a layer of foam is that:
   A. Fire needs air to continue burning.
   B. The layer of foam will allow the heat to escape.
   C. The ignition temperature of the substance will be lowered by the foam.
   D. Heat from the fire will break down the carbon dioxide.

3. The reason that foam is 90% carbon dioxide by volume and 85% water by weight is that:
   A. Carbon dioxide is a pure gas.
   B. Carbon dioxide weighs more than air.
   C. Carbon dioxide occupies less volume than water.
   D. Carbon dioxide has less density than water.

4. Foam fire extinguisher are not effective on fires involving:
   A. Ignited wood.
   B. Ignited fabric.
   C. Ignited gasoline.
   D. Ignited alcohol.

5. Foam fire extinguisher expel their ingredients by the use of:
   A. The force created by the stabilizer acting in the solution.
   B. A Firefighters operating the hand-pump.
   C. The gas pressure created by the chemical reaction.
   D. The expanding of the water in the inside section.

ANSWERS:

1. C   2. A   3. D   4. D   5. C

STATEMENT

Selection of the best portable fire extinguisher for a given situation depends on:

1. The nature of the combustibles which might be ignited.
2. The potential severity (size, intensity, and speed of travel) of any resulting fire.
3. The effectiveness of the extinguisher on that hazard.
4. The ease of use of the extinguisher.
5. The personnel available to operate the extinguisher and their experience and training along with emotional reactions as influenced by their training.
6. The ambient temperature conditions and other special atmospheric considerations (wind, draft, presence of fumes).
7. The suitability of the extinguisher for its environment.
8. Any anticipated adverse chemical reactions between the extinguishing agent and the burning material.
9. Any health and operational safety concerns (exposures of operators during the fire control)
10. The upkeep and maintenance requirements for the extinguisher.

Portable fire extinguisher are designed to cope with fires of limited size and are necessary and desirable even though the property may be equipped with automatic sprinkler protection, standpipe and hose systems, or other fixed fire protective equipment.

The initial selection of the type and capacity of an extinguisher is based on the hazards of the area to be protected. NFPA #10, Standard for the Installation, Maintenance, and use of Portable Fire Extinguisher, i.e. "NFPA Extinguisher Standard", has established three hazard levels in order to provide a simplified method of determining the portable size of a fire relative to the kind of incipient fire and its potential severity.

Light Hazard:

Where the amount of combustibles or flammable liquids present is such that fires of small size may be expected. These may include offices, schoolrooms, churches, assembly halls, telephone exchanges, etc.

Ordinary Hazards:

Where the amount of combustibles or flammable liquids present is such that fires of moderate size may be expected. These may include mercantile storage and display areas, auto showrooms, parking garages, light manufacturing areas, warehouses not classified as extra hazard, school shop areas, etc.

STATEMENT
(continued)

Extra Hazards:

Where the amount of combustibles or flammable liquids present is such that fires of severe magnitude may be expected. These may include woodworking areas, auto repair shops, aircraft servicing areas, warehouses with high-piled combustibles (over 15 feet in solid piles, over 12 feet in piles that contain horizontal channels), and areas involved with processes such as flammable liquid handling, painting, dipping, etc.

The class of hazard can influence the type of extinguisher selected as well as the size of fire extinguishing capability (i.e., 2 1/2 gallon capacity stored pressure or pump tank water extinguisher are rated 2-A and are only suitable for light or ordinary hazard protection. When extra hazard conditions exist, multipurpose dry chemical extinguisher having ratings of 3-A to 40-A will provide the degree of protection needed).

ACCORDING TO THE PREVIOUS STATEMENT

1. Selection of the best portable fire extinguisher for a given situation depends on:
    A. Nature of combustibles that are ignited.
    B. Potential severity of any resulting fire.
    C. Ease of use.
    D. All of the above.

2. Portable fire extinguisher are designed to cope with fires:
    A. Of limited size.
    B. Of any size.
    C. In outside areas only.
    D. In inside areas only.

3. The initial selection of the type and capacity of an extinguisher is based upon:
    A. The size of the fire.
    B. The location of the fire.
    C. The hazard of area to be protected.
    D. The hazard of the contents of extinguisher.

4. The NFPA Standard for the Installation, Maintenance and use of Portable Fire Extinguisher is:
    A. NFPA #8.
    B. NFPA #9.
    C. NFPA #10.
    D. NFPA #15.

5. Light Hazard occupancies include all of the following, EXCEPT:
    A. Offices.
    B. Schoolrooms.
    C. Hospitals.
    D. Assembly halls.

6. Ordinary Hazard occupancies include all of the following, EXCEPT:
    A. Mercantile storage.
    B. Auto repair shops.
    C. Auto showrooms.
    D. Parking garages.

7. Extra Hazard occupancies include all of the following, EXCEPT:
    A. Aircraft servicing areas.
    B. Auto painting shops.
    C. Warehouses with high-piled combustibles.
    D. School auditoriums.

8. When Extra Hazard conditions exist, it is recommended that: which of the following ratings be provided?
    A. 2 1/2 gallon water.
    B. Dry chemical, 2-A to 15-A.
    C. Dry chemical, 3-A to 40-A.
    D. Multipurpose dry chemical, 3-A to 40-A.

ANSWERS:

| 1. D | 3. C | 5. C | 7. D |
| 2. A | 4. C | 6. B | 8. D |

STATEMENT

In general, liquefied gas extinguisher: bromotrifluoromethane (halon 1301) and bromochlorodifluoromethane (halon 1211), have features and characteristics similar to CO-2 extinguisher. The bromotrifluromethane (halon 1301) extinguisher has never been available in a size larger than 2 1/2 lbs. It has a listed rating of 2-B:C which is below the minimum requirements. This extinguisher does not appear to have a large advantage over other liquified gas extinguisher.

The bromochlorodifluromethane (halon 1211) extinguisher is available in a wide range of sizes, with listed ratings of 2-B:C to 10-B:C. The agent is similar to CO-2 in that it is suitable for cold weather installation, is noncorrosive, and leaves no residue. It is considerably more effective on small Class A fires than CO-2; however water may still be needed as a follow-up to extinguish glowing embers and deep-seeded burning. Compared to CO-2 on a weight-of-agent basis, bromochlorodifluromethane (halon 1211) is at least twice as effective. When discharged, the agent is in the combined form of a gas/mist with about twice the range of CO-2. To some extent, windy conditions or strong air currents may make extinguishment difficult by causing the rapid dispersal of the agent. The shell for the halon 1211 extinguisher are light weight aluminum or mild-steel and weigh considerably less than CO-2 cylinders.

ACCORDING TO THE PREVIOUS STATEMENT

1. Liquified gas extinguisher such as halon 1301 and halon 1211 have features and characteristic similar to:
    A. CO-2 extinguisher.
    B. 2-A 10 BC extinguisher.
    C. Dry chemical extinguisher.
    D. Multipurpose dry chemical extinguisher.

2. Halon 1211 extinguisher are not available in sizes above:
    A. 2 1/2 lbs.
    B. 3 lbs.
    C. 5 lbs.
    D. None of the above.

3. Halon 1301 extinguisher are not available in sizes above:
    A. 2 1/2 lbs.
    B. 3 lbs.
    C. 5 lbs.
    D. None of the above.

4. Halon 1211 is:
    A. Not suitable for cold weather installation.
    B. Not similar to CO-2.
    C. Noncorrosive.
    D. The one that leaves a residue.

5. Halon 1211 as compared to CO-2 on a weight-to-agent basis, is:
    A. About 1/2 as effective.
    B. About 2 times as effective.
    C. Has about 1/2 the range.
    D. Uses a heavier container.

ANSWERS:

1. A   2. D   3. A   4. C   5. B

STATEMENT

Fire strategy is the plan of attack on a fire. Fire strategy should make prime use of equipment and personnel and take into consideration fire behavior, the nature of the occupancy, environmental conditions, and weather factors. Fire tactics are the various maneuvers that can be employed in a strategy to successfully fight a fire. Even though no two fires are alike, it is possible to lay down general plans for firefighting operations primarily because the elements of similarity are sufficient enough to establish fire strategy and fire tactics that are applicable in a variety of situations. In any fire remember for covering all points of a fire, cover the front and rear, and over and under along with covering all exposures.

ACCORDING TO THE PREVIOUS STATEMENT

1. Fire strategy is:
    A. The attack on a fire.
    B. The evaluation of a fire scene.
    C. The plan of attack on a fire.
    D. The critique of a fire.

2. Fire tactics are:
    A. The extinguishment of a fire.
    B. The critique of a fire.
    C. The various apparatus and equipment that may be employed in a strategy to successfully fight a fire.
    D. The various maneuvers that can be employed in a strategy to successfully fight a fire.

3. Fire strategy should:
    A. Make prime use of equipment and personnel.
    B. Take fire behavior into consideration.
    C. Take the weather into consideration.
    D. All of the above.

4. It is possible to lay down general plans for firefighting, because:
    A. No two fires are alike.
    B. All fires are alike.
    C. The elements of similarity are sufficient enough to establish tactics and strategy.
    D. The elements of tactics and strategy are always exactly the same.

5. In covering all points of a fire officers should remember to cover:
    A. The front and rear.
    B. Over and under.
    C. All exposures.
    D. All of the above.

ANSWERS:

1. C   2. D   3. D   4. C   5. D

STATEMENT

When a superior officer takes over the command of a large fire from a subordinate officer that has been in charge for some time, he should make a review the actions taken up to this time. The first action that an officer should check when relieving another officer at a fire scene, is whether the fire is adequately surrounded by the apparatus and Firefighters available. If the structure is fully involved in fire, which is free of exposures on all sides, it is normally best to direct water streams to the structure from three sides rather than from all four sides, mainly because the heat, smoke and gases will be driven away from the Firefighters, permitting them to work more proficiently. Remember that the most common mistake that the first in officer may have made is not to have requested the proper additional help needed, because manpower is the most critical at the early stages of a fire.

ACCORDING TO THE PREVIOUS STATEMENT

1. When a superior officer takes over a fire from a subordinate officer:
    A. Make a check of the actions taken up to this time.
    B. Cancel all previous orders.
    C. Call for additional help.
    D. Make a list of previous improper actions.

2. When an officer relieves another officer at a fire, his FIRST action taken should be:
    A. Check for adequate personnel and apparatus.
    B. Call the media.
    C. Check to see if the fire is adequately surrounded by apparatus and Firefighters.
    D. See if the fire was deliberately started.

3. When a structure is fully involved in fire, and is free on all sides, it is best to direct water streams to the structure from:
    A. The front.
    B. The rear.
    C. Three sides.
    D. All four sides.

4. The most common mistake that the first in officer will make is:
    A. Deliver too much water.
    B. Neglect to call for additional help.
    C. Request too much help.
    D. Attack the fire too aggressively.

5. Manpower: most critical at what stage of fire?
    A. Early stages.
    B. Middle stages.
    C. Late stages.
    D. None of the above.

ANSWERS:

1. A   2. C   3. C   4. B   5. A

STATEMENT

Centrifugal pumps employ a certain principle of force in pumping. The power or force to create pressure is exerted from the center. The revolving motion of the impeller will whirl water introduced at the center toward the outer edge of the impeller. Here it is trapped by the pump casing and is forced to the discharge outlet. In the centrifugal pump the power is transmitted from the drive shaft, through the pump transmission, and intermediate gear to the impeller shaft. With the quantity remaining constant in a centrifugal pump the pressure will increase at a rate equal to the square of the speed increase. The pump speed is greater than the engine speed in a centrifugal pump. Centrifugal pumps cannot create a vacuum.

ACCORDING TO THE PREVIOUS STATEMENT

1. The force or power in centrifugal pumps is exerted from:
    A. The outside edge of the of the impeller.
    B. The center of the pump.
    C. The discharge outlet.
    D. The intake valve.

2. Centrifugal pumps power is transmitted FROM:
    A. The drive shaft.
    B. The pump transmission.
    C. The intermediate gear.
    D. The impeller.

3. In centrifugal pumps the water is introduced at:
    A. Outer edge of the impeller.
    B. Center towards the outer edge.
    C. Outer edge towards the center.
    D. The pump casing.

4. Centrifugal pumps are NOT capable of creating:
    A. Pressure.
    B. Force.
    C. Vacuum.
    D. Discharge.

5. In centrifugal pumps, with the quantity remaining constant, the pressure will increase at a rate equal to:
    A. The square of the quantity increase.
    B. The square of the volume increase.
    C. The square of the discharge increase.
    D. The square of the speed increase.

ANSWERS:

1. B   2. A   3. B   4. C   5. D

STATEMENT

Water under pressure, in the volute of a centrifugal pump is prevented from returning to the suction side of the pump by the close fit of the impeller hub to a stationary wear ring at the eye of the impeller, and hydraulic pressure due to the velocity created by the centrifugal force. The volute enables the centrifugal pump to handle the increasing quantity of water towards the discharging outlet and at the same time permit the velocity of the water to remain constant or to decrease gradually maintaining the continuity of flow. The volute is a progressively expanding waterway which converts velocity to pressure as the velocity remains constant. The volute principle is the design of the water passageway in centrifugal pumps.

ACCORDING TO THE PREVIOUS STATEMENT

1. In the volute of a centrifugal pump, the water under pressure is prevented from returning to the suction side of the pump by:
    A. Close fit of the impeller.
    B. Hydraulic pressure created in the pump.
    C. Both A and B.
    D. None of the above.

2. Centrifugal pumps volute enables the pump to:
    A. Handle the increasing quantity of water towards the discharge outlet.
    B. Permit the velocity of the water to remain constant.
    C. Permit the velocity of the water to decrease.
    D. All of the above.

3. The volute of a centrifugal pump converts:
    A. Pressure to velocity.
    B. Velocity to pressure.
    C. Pressure to volume.
    D. Volume to pressure.

4. The volute principle in a centrifugal pump is the design of:
    A. The airway.
    B. The impeller.
    C. Discharge valve.
    D. The water passageway.

5. The volute in a centrifugal pump is a:
    A. Decreasing waterway.
    B. Expanding waterway.
    C. Constant waterway.
    D. Unchanging waterway.

ANSWERS:

1. C   2. D   3. B   4. D   5. B

STATEMENT

The present recognized capacities of pumps for Fire Department pumpers are: 500, 700, 1000, 1250, and 1500 gallons per minute (GPM), although some larger capacity pumpers have been built. In order for a Fire Department pumper to meet standard requirements, it must deliver its rated GPM capacity at 150 pounds per square inch (PSI). The pumper further must deliver 70% of its rated capacity at 200 PSI and, 50% of its rated capacity at 250 PSI net pump pressure. A Fire Department pumper must be provided with adequate inlet and discharge pump connections, pump and engine controls, gauges, and other instruments.

Some Fire Department pumpers are referred to as a "Triple Combination Pumper". To qualify as a triple combination pumper, a pumper must have a fire pump, a hose compartment, and a water tank. In addition, a Fire Department pumper will usually carry ladders, tools and equipment, hose, and other accessories. The location of the three components of a triple combination pumper may vary with each manufacture's design and specifications as written by the purchaser.

The main purpose of a Fire Department pumper is to provide adequate pressure for fire streams. The water it pumps may come from its water tank, a fire hydrant, or an impounded supply.

ACCORDING TO THE PREVIOUS STATEMENT

1. Of the following, which is NOT a recognized capacity of pump for Fire Department pumpers:
    A. 1000 GPM.
    B. 1250 GPM.
    C. 1500 GPM.
    D. 1750 GPM.

2. Fire Department pumpers must deliver 100% of their rated capacity at a pressure of:
    A. 100 PSI.
    B. 150 PSI.
    C. 200 PSI.
    D. 250 PSI.

3. Fire Department pumpers must deliver 70% of their rated capacity at a pressure of:
    A. 100 PSI.
    B. 150 PSI.
    C. 200 PSI.
    D. 250 PSI.

4. Fire Department pumpers must deliver 50% of their rated capacity at a pressure of:
    A. 100 PSI.
    B. 150 PSI.
    C. 200 PSI.
    D. 250 PSI.

5. Fire pumpers must be provided with adequate:
    A. Inlet and discharge pump connections.
    B. Pump and engine controls.
    C. Gauges and other instruments.
    D. All of the above.

6. A Fire Department triple combination pumper must have all of the following, EXCEPT:
    A. Fire pump.
    B. Water tank.
    C. Hose compartment.
    D. Hose.

7. The main purpose of a Fire pumper is to:
    A. Deliver Firefighters to fire incidents.
    B. Provide adequate pressure for fire streams.
    C. Transport tools and equipment.
    D. Transport water.

ANSWERS:

1. D   2. B   3. C   4. D   5. D   6. D   7. B

STATEMENT

The process of raising a ladder where it is needed will not in itself extinguish fire, but a well-positioned ladder becomes a means by which other operations can be performed. Fire Service ladders are essential for a rescue procedure, and carrying firefighting tools and appliances need to be carried above ground level. Teamwork, smoothness, and rhythm are necessary when raising and lowering Fire Department ladders, if speed and accuracy are to be developed. A knowledge of how ladders are mounted on fire apparatus, how they can be removed and used, along with an understanding of safety, spacing, and climbing techniques, is essential. These preliminary methods and skills provide a background for handling ladders. The recognized specifications for ladders recommend that they be made of metal.

ACCORDING TO THE PREVIOUS STATEMENT

1. Raising a ladder where it is needed in itself:
    A. Will extinguish a fire.
    B. Allow other operations to be performed.
    C. Both A and B.
    D. none of the above.

2. Ladders are essential for all of the following, EXCEPT:
    A. Rescue procedures.
    B. Delivering tools and equipment above ground level.
    C. In many Fire Department operations.
    D. In all Fire Department operations.

3. When raising and lowering ladders, it is necessary to have:
    A. Teamwork.
    B. Rhythm.
    C. Smoothness
    D. All of the above.

4. Preliminary methods and skills that provide the knowledge and background for handling ladders include all of the following, EXCEPT:
    A. How ladders are mounted on apparatus.
    B. How ladders may be removed from apparatus.
    C. Where ladders are purchased.
    D. Understanding safety, spacing, and climbing techniques.

5. Recognized specifications for ladders recommend that Fire Department ladders be made of:
    A. Birch.
    B. Oak.
    C. Ash.
    D. Metal.

ANSWERS:

1. B   2. D   3. D   4. C   5. D

STATEMENT

The fire triangle is a three sided figure that represents three factors necessary for combustion. The three factors necessary for combustion are: oxygen, heat, and fuel. The fire tetrahedron represents the four elements that are required by fire. The four elements required by a fire are: fuel, heat, oxygen, and uninhibited chain reaction.

ACCORDING TO THE PREVIOUS STATEMENT

1. The fire triangle is a three sided figure that represents three factors necessary for:
    A. Fire extinguishment.
    B. Fire safety.
    C. Fire combustion.
    D. Fire prevention.

2. The fire triangle is a three sided figure that represents the three factors that are necessary for combustion, which of the following is not one of these factors.
    A. Oxygen.
    B. Carbon monoxide.
    C. Heat.
    D. Fuel.

3. What does the fire tetrahedron represent?
    A. The three factors required for combustion.
    B. The three factors required for fire extinguishment.
    C. The four factors required for fire extinguishment.
    D. The four elements required by fire.

4. All but one of the following are elements required by a fire, which one is it?
    A. Fuel.
    B. Heat.
    C. Inhibited chain reaction.
    D. Uninhibited chain reaction.
    E. Oxygen.

5. One of the following is not a factor of the fire triangle, which one is it?
    A. Fuel.
    B. Heat.
    C. Oxygen.
    D. Inhibited chain reaction.

ANSWERS:

1. C   2. B   3. D   4. C   5. D

SUMMARY

In this chapter we have gained experience by reading several typical exam type reading questions. This chapter has given you needed practice and experience in these types of questions. You can expect to see these types of questions on entrance level Firefighter exams.

Remember practice makes perfect. Practice is best done for a short time every day rather than long periods once or twice a week.

CHAPTER 9

BASIC SCIENCE: CHEMISTRY & PHYSICS

BASIC SCIENCE/CHEMISTRY/PHYSICS

INTRODUCTION

Entrance level Firefighter exams will usually have some questions relating to Science. The Science questions will usually cover a broad range of subjects pertaining to General Science, Fire Science, Fire Chemistry, and Physics of Fire.

When taking an examination and you encounter Science, Chemistry, or Physics type questions: don't be afraid if your knowledge is not very broad. General knowledge of the basic concepts will be adequate.

If you have knowledge of any foreign languages, use this knowledge when recalling the meanings of technical terms. For example, Latin roots of words are particularly revealing.

Use your common sense when dealing with unfamiliar areas. What seems to be the most sensible answer will normally be the correct answer.

SCIENCE: CHEMISTRY/PHYSICS

INFORMATION RELATING TO THE FIRE SERVICE:

B.T.U. = British thermal unit.

B.T.U. = 1 LB water increases 1 degree F.

B.T.U. is the amount of heat required to raise one pound of water one degree F. (at atmospheric pressure).

To convert one pound of ice at 32 degrees F to steam at 212 degrees F requires **1293.7 B.T.U.'s.**

To convert ice to water requires **143.4 B.T.U.'s.**

To convert water to steam requires **970.3 B.T.U.'s.**

To raise 32 degrees F to 212 degrees F requires **180 B.T.U.'s.**

One pound of construction wood = **8120 B.T.U.'s.**

One gallon of water will absorb about **8000 B.T.U.'s.**

SPECIFIC HEAT = number of B.T.U.'s to raise 1 LB of water 1 degree F.

SPECIFIC HEAT = absorption of heat. Water has a high specific rating.

LATENT HEAT = absorbed heat or heat given off.

LATENT HEAT = heat absorbed or given off from a substance as it transfers from a liquid to a gas or a solid to a liquid.

SUBLIME: is when a solid changes to vapor without passing through the liquid phase.

VAPOR = substance in a gaseous state, particularly that of liquid or solid at normal temperatures.

When converting liquid to vapor, the **VOLUME IS INCREASED BY 1700 TIMES.**

A gallon of water may produce a maximum of **200 CUBIC FEET OF STEAM.**

VAPOR DENSITY = ratio of gases.

VAPOR DENSITY = the weight of a determined volume of a gas or vapor, as related to the weight of the same volume of normal dry air.

VAPOR DENSITY of air is = 1, gases of less than 1 the gas is lighter than air. Gases that have a rating of more than 1 are heavier than air.

The **RATE OF DIFFUSION** of an unconfined gas varies inversely with its **VAPOR DENSITY.**

SPECIFIC GRAVITY = ratio of weight to volume.

SPECIFIC GRAVITY = the comparison of the weight or mass of a given volume of a substance at a specific temperature to that of an equal volume of water.

SPECIFIC GRAVITY applies to liquids only.

SPECIFIC GRAVITY is the ratio of solid weight or liquid to volume of water. Water = 1 at 4 degrees C or 39 degrees F.

MAXIMUM DENSITY of water is at **39 DEGREES F.**

Liquids of a specific gravity of less than 1 will **FLOAT ON WATER**.

Liquids of specific gravity of more than 1, **WATER WILL FLOAT ON TOP** of this liquid.

LIQUIDS will not expand indefinitely like gases.

A **SUBSTANCE** = gas if it is in a gaseous state at 100 degrees F and 40 PSI.

A **SUBSTANCE** = liquid if it is in a liquid state at 70 degrees F and 14.7 PSI.

BOILING POINT = temperature when vapor is equal to atmospheric pressure.

BOILING POINT = the temperature that a liquid will swiftly convert to vapor.

Most **USEFUL** information concerning hazard of a liquid is the **FLASH POINT**.

FLASH POINT more than any other physical property determines the hazard of a liquid.

FLASH POINT = the lowest temperature of liquid at which it gives off vapor which forms ignitible mixture with air at the surface of the liquid or within the vessel it is used. (capable of propagation of flame with heat, fire does not continue with a heat source).

FIRE POINT = temperature that a **LIQUID** is able to continue to burn without an outside heat source.

FIRE POINT is the lowest temperature of a **LIQUID** at which vapors are evolved fast enough to continue combustion.

IGNITION TEMPERATURE = temperature that a **SUBSTANCE** is able to continue to burn without an outside heat source.

IGNITION TEMPERATURE is the temperature at which a **SOLID COMBUSTIBLE SUBSTANCE** will ignite and burn.

IGNITION TEMPERATURE of combustible natural gas-air mixture is 1000 degrees F.

FLAMMABLE DENSITY = the range of combustible vapors or gas mixtures with air between the upper and lower flammable limits.

FLAMMABLE LIQUIDS are liquids with flash points **BELOW 100 DEGREES F.**

COMBUSTIBLE LIQUIDS are liquids with flash points **AT 100 DEGREES OR HIGHER.**

Combustible liquids are **SAFER** than flammable liquids.

VAPOR PRESSURE OF A LIQUID: is the pressure of the vapor at any given temperature at which the vapor and liquid phases of the substance are in balance in a closed vessel.

Generally as the **BOILING POINT** of a liquid goes down, the **VAPOR PRESSURE** and evaporation rat **INCREASE**.

FOR THE FOLLOWING TWO CHARTS :
    FP = FLASH POINT AND BP = BOILING POINT.

CLASSIFICATION OF FLAMMABLE LIQUIDS:

Class I   have FP less than 73 Degrees F.
Class IA have FP less than 73 degrees F.; BP less than 100 degrees F
Class IB have FP less than 73 degrees F.; BP greater than 100 degrees F.
Class IC have FP between 73 and 99 Degrees F.

CLASSIFICATION OF COMBUSTIBLE LIQUIDS:

Class II have   FP between 100 and 140 degrees F.
Class III have FP above   140 degrees F.
Class IIIA have FP between 140 and 199 degrees F.
Class IIIB have FP above   200 degrees F.

VAPORIZATION: is the process that a substance changes from liquid or solid phase to a gas.

1 degree C = 1.8 degrees F - (+32 degrees F). Example: 50C = 90F + 32F = 122 degrees F. **THIS IS FOR POSITIVE TEMPERATURES!**

1 degree C = 1.8 degrees F - (+32 degrees). Example: -40C = 72F -(+32F) = 40 degrees F. **THIS IS FOR NEGATIVE TEMPERATURES!**

-40 degrees F and -40 degrees C is the only temperature that degrees F and degrees C are **IDENTICAL**.

Steel sparks = **2500 DEGREES F.**

Temperature of a match flame = **2000 DEGREE F.**

Temperature of a cigarette = **550 - 1350 DEGREES F.**

SALT WATER boils at **226 DEGREES F.**

Unprotected steel looses its strength at about **1000 DEGREES F.**

Bare steel has a fire resistance of about **TEN MINUTES**.

Under severe fire conditions, columns of cast iron and unprotected steel collapse in **10 - 20 MINUTES**.

Firefighters should be aware that most metals **EXPAND WHEN HEATED**.

Wood burns at a rate of **1" IN 45 MINUTES**.

The maximum temperature that wood should be exposed to constantly is **300 DEGREES F.**

Cheapest construction material is **WINDOW GLASS**.

Fires usually burn **UPWARD AND OUTWARD**.

In most structure fires, the floor temperatures are usually about **ONE THIRD** that of the ceiling temperatures.

SPONTANEOUS is slow oxidation.

CHEMICAL REACTIONS double their rate with each **18%** rise in temperature.

In this country the **MAJORITY** of fires occur in dwellings confined to **ONE ROOM**.

The leading cause of fire is **SMOKING**.

Dwelling fires usually involve the **KITCHEN**.

ROUGH DARK SURFACES are the best surfaces for absorbing and radiating heat.

Axe handles are usually made of wood rather that metal primarily because the wooden handles will cushion the impact to the Firefighter.

Ice on sidewalks will melt from the application of salt by the lowering of the freezing point of the water by the salt.

The main hazard of static electricity is the possibility of an explosion created by sparks.

Usually the color and odor of smoke will indicate the kind of material that is burning.
In warm climate areas, water temperature may be lowered by storing the water in earthenware containers, because some of the water changes to vapor, which lowers the water temperature.

A wet hand will stick to a piece of metal but not to a piece of wood, at certain temperatures, because metal is a better conductor of temperature.

Substances that are subject to spontaneous combustion are capable of igniting without an external source of heat.

The air that humans breath out has more water vapor in it than the air that humans breath in.

If the formation of ice takes place within a closed container, the container will burst because the water expands when it freezes, which builds up great pressure.

Kerosene lamps burn with a yellow flame because of the incandescence of unburned carbon particles.

When metal pipe is used to carry liquids or gases, the threaded portion is always tapered.

Smoke usually rises from a fire because the cooler, heavier air displaces the lighter warm air.

Water heated at sea level will boil at a higher temperature than water heated at high altitudes.

Good conductors of heat are usually poor insulators of heat.

Chemically pure water may be made from tap water by the process of distillation.

When air rises it cools because it expands.

A catalyst is a substance that will change the speed of a chemical reaction.

The sprocket wheels and chain of a bicycle increase the speed of the rear wheel. The height at which an object will float on water is determined mainly by the weight of the water displaced by the object.

FIRE HYDRAULICS

WATER weighs approximately **62.5** LBS per cubic foot.

One cubic foot of **WATER** contains **1728** cubic inches.

One gallon of **WATER** contains **231** cubic inches.

One gallon of **WATER** weighs **8.35** LBS.

There are **7.481** gallons in a cubic foot.

A column of **WATER** 1" high and 1" square weighs **.434** LBS. (note .5 PSI for field calculations).

A column of **WATER 2.304** feet in height, will exert **1** LB PSI at its base.

Four **FUNDAMENTALS** that govern friction loss in hose lines and pipes:
    1. All other conditions being equal the loss by friction varies with the length of the line.
    2. In the same size hose, friction loss varies directly as the square of velocity flow.
    3. For the same flow, friction loss varies inversely as the fifth power of the diameter of the hose.
    4. For a given velocity flow, friction loss is independent of the pressure.

Friction loss in two lines of 2 1/2" hose is about **28%** of the friction loss in a single line of 2 1/2" hose.

Friction loss in old hose may be **50%** greater than new hose.

Friction loss is governed by the **QUANTITY (GPM)** of water flowing.

Hose in a **ZIG-ZAG** pattern will increase friction loss by about **6%**.

If the length of hose is **DOUBLED** the friction loss is **DOUBLED**.

Pressure **DOES NOT** effect friction loss in the same size hose. (increase or decreases)

MERCURY weighs **.49 LBS** per cubic inch.
>   **EXAMPLE:**
>   30 Hg" indicates 30 X .49 = 14.7 LBS atmospheric pressure.

NORMAL ATMOSPHERIC PRESSURE:
>   29.92 Hg" or 14.7 PSI; (29.92 X .49 = 14.7)

PERFECT VACUUM; water could theoretically allow atmospheric pressure to raise water 33.9 feet. (29.92 Hg" X 1.13 = 33.92)

ATMOSPHERIC PRESSURE of 14.7 PSI is capable of sustaining a column of water 33.9 feet high (14.7 X 2.304 = 33.9), this is if a pump could produce a perfect vacuum. (the practical limit is about 22 feet).

ATMOSPHERIC PRESSURE decreases as elevation increases at a rate of approximately 1/2 LB every 1000 feet of elevation.

PRESSURE may be defined as the measurement of energy contained in water.

The **THEORETICAL HEIGHT** to which a pumper can raise water by suction decreases proportionally as elevation increases, therefore the theoretical height to which a pumper can raise water decreases about 1 foot every 1000 feet of altitude.

PRESSURE, as we normally think of it in the fire service is that force delivered by pumpers for supplying water to a fire for firefighting.

BACK PRESSURE = .434 X height. (for field use .5 PSI) or 1/2 of the height = head.

PRESSURE: force per unit area measured in PSI.

BACK PRESSURE: the pressure from a static head of water.

Normally **PRESSURE** works for us, but when pumping up hills or to nozzles located on upper floors of buildings, it can work against us. (back pressure).

HEAD is the vertical distance from the surface of the water being considered to the point being considered.

If the nozzle is to be lower than the pump, the factor of elevation must be **SUBTRACTED** from the pump pressure.

NOZZLE PRESSURE on a charged line with nozzle closed is approximately equal to the same pressure as the engine pressure. (+ or - the head)

NOZZLE PRESSURE: the velocity pressure in PSI at which water is discharged from a nozzle.

NOZZLE REACTION is caused by an increase in water velocity.

VELOCITY FLOW is the speed at which water passes a given point.

Pressure has **NO EFFECT** on the friction loss as long as it does not change the velocity of the flow. It is the **VELOCITY** of the water over the lining of the hose and the condition of the lining that governs the friction loss.

PHYSICS

PHYSICS is the science that deals with energy and matter.

DYNAMICS is the general term used to describe the laws that govern forces in which motion is produced.

CENTRIFUGAL FORCE is the tendency of an object traveling in a curve to try and go in a straight line.

GRAVITATION is the physical occurrence of weight that an object maintains.

GRAVITY is the term used to describe the attraction that exist between the earth and the bodies on or near it.

WEIGHT is the measure of the earths attraction for a mass.

The loss of weight of an object submerged in a liquid is equal to the weight of the displaced liquid.

Scissors are an example of a first class **LEVER**.

It is easier to turn a wrench with a long handle than a short handle, because it provides more **LEVERAGE**.

FULCRUM is the fixed point on which a lever turns when it is used.

The **SCREW** is a powerful machine capable of moving large objects.

The most common way to determine **SPECIFIC GRAVITY** is with the use of a **HYDROMETER**.

FRICTION is the force which tends to keep an object, placed on an incline, from slipping down.

FRICTION is the resistance encountered when an object moves over another object.

FLUID FRICTION depends upon the speed of the flow.

The type of electricity produced by **FRICTION** is called **STATIC ELECTRICITY**.

Machinery is oiled in order to decrease **FRICTION** of moving parts.

For every **ACTION** there is a **REACTION**.

NEWTON discovered the law of inertia.

HOOKE'S LAW states that strain is proportional to stress.

RADIUS is the distance from the center to the edge of a circle or a sphere.

CIRCUMFERENCE is the line around a circle.

ANGLE is the space between two lines or surfaces that meet at a point or that cross each other.

ACUTE ANGLE is an angle that is smaller than a right angle, less than 90 degrees.

RIGHT ANGLE is an angle which is at 90 degrees.

OBTUSE ANGLE is an angle that is greater than a right angle, more than 90 degrees.

ANGLES are measured in terms of degrees.

SHEAR is a type of stress.

When a **MAGNET** is set into a mound of iron shavings, the shavings will cling near the ends.

The most common **MAGNETIC** substance is **IRON**.

ARMATURE is an iron core that will rotate within pole pieces.

The principle parts of an electric motor are a stationary field **ELECTROMAGNET** and an **ARMATURE MAGNET**.

A permanent **MAGNET** is used in a speedometer.

DYNAMO is a machine that converts electrical energy into mechanical energy.

Silver is one of the metal substances that **MAGNETIZATION** is least easily produced.

TURBINE is a type of engine operated by the pressure of water, steam, or air on ridges or fins, called vanes, on the side of a rotating disk.

The principle for the operation of a **STEAM ENGINE** is based on steam compression.

The capacity of a **STEAM BOILER** is the amount of steam it can produce in an **HOUR**.

BIMETALLIC THERMOSTAT function because of the fact that different metals expand at different rates when heated.

CONDENSER is able to stop direct currents.

PORCELAIN is a very good electrical insulator.

AMPERES is the term used to describe the strength of electrical current.
CONDUCTOR is any material which will allow an electric current to pass through it.

ALTERNATING CURRENT is the electrical current that reverses its polarity at rapid regular intervals.

TRANSFORMER will change the voltage of an alternating electrical current.

INSULATOR is a substance that offers high resistance to the passage of electricity.

PRESSURE is the instrument which causes electrical current to flow through wire.

BAROMETER is an instrument that measures atmospheric pressure.

When water freezes it **EXPANDS**.

SUCTION is created by **VACUUM**.

Liquid will rise into a straw, into your mouth, and upwards because the pressure on the liquid is greater than the pressure in the mouth.

BAROMETER is the instrument that is used to determine the pressure of air changes.

ELECTRON is a particle of negative electricity.

PROTONS are the positive particles of electricity contained within atoms.

ELECTROLYSIS is the process of decomposing water by electricity.

HEAT will make the molecules in ice move faster.

FUEL is whatever is burned to create **HEAT**.

CALORIE is the unit of **HEAT**.

HEAT and **LIGHT** pass more than ninety millions of miles from the sun to the earth by radiation. With the **LIGHT** traveling from the sun to the earth in about eight seconds.

HEAT passes through iron by conduction.

Metal placed into a hot liquid will **EXPAND** because of conduction.

ELEMENTS contain heat energy, and produce heat while they are uniting.

SOUND travels at about 1090 feet per second.

GYROSCOPE is a device that is used to increase stability.

FOOT POUND is the unit applied to **WORK**.

ADIABATIC is a change in the condition of a material that occurs without gain or loss of heat from surrounding material.

KINETIC PRESSURE is moving pressure or pressure in motion, such as water moving through hose lines under pressure.

SUMMARY

In this chapter it was pointed out that entrance level Firefighter exams will usually have some questions relating to Science. The Science questions will usually cover a broad range of subjects pertaining to General Science, Fire Science, Fire Chemistry, and Physics of Fire.

Remember when taking an examination and you encounter Science, Chemistry, or Physics type questions: don't be afraid if your knowledge is not very broad. General knowledge of the basic concepts will be adequate. If you find it necessary to improve your knowledge in these areas, you might consider taking a class in basic chemistry and/or physics.

It is apparent that you that you will have to be familiar with the terms discussed in this session and why these terms are important to Firefighters. If you have knowledge of any foreign languages, use this knowledge when recalling the meanings of technical terms. For example, Latin roots of words are particularly revealing.

ASSIGNMENT:

1. Read: "Forcible Entry - Salvage - Overhaul - Ventilation": Information Sheet #11. Pages 364 - 365.

DIAGNOSTIC TEST

CHEMISTRY AND PHYSICS

REMEMBER: read the question and all answer choices carefully prior to determining the correct answer. Go on to the next question if you cannot select an answer quickly. Actual exams are timed.

1. Which of the following statements is correct:

   A. Water heated at sea level will boil at a higher temperature than water heated on the top of a mountain.
   B. A large quantity of water will boil at a higher temperature than a small quantity.
   C. Water heated slowly by a low flame will boil at a higher temperature than water heated quickly by a high flame.
   D. Water always boils at the same temperature regardless of pressure.

2. Earthenware jugs will keep water cool, even in hot temperatures, because:

   A. Particles of earthenware dissolve in the water and lower the waters temperature.
   B. The rough surface of the jug radiates heat more rapidly.
   C. The change of some of the water to a vapor lowers the temperature.
   D. The jug is a good conductor of heat.

3. Of the following toxic gases, the one which is the most dangerous because it cannot be seen and has no odor is:

   A. Ammonia.
   B. Chlorine.
   C. Carbon monoxide.
   D. Ether.

4. If oil poured on water it will reduce the height of the waves, because:

   A. Oil fills up the toughs of the waves.
   B. Chemical action between oil and water produces a heavier substance.
   C. Added weight of the oil makes the waves break at a lesser height.
   D. Surface tension of the water will be weakened.

5. When kerosene burns with a yellow flame, this is due to the:

   A. Burning of hydrogen.
   B. Incandescence of unburned carbon particles.
   C. Complete burning of the hydrocarbons.
   D. Heating of the wick.

6. The threads on metal pipe that is used for liquids and gases will always be:

   A. Straight.
   B. Fine.
   C. Coarse.
   D. Tapered.

7. The following principle that applies to a jet plane is:

   A. Energy can be neither created or destroyed.
   B. If pressure is applied to a confined gas, the volume of the gas will decrease.
   C. For every action there is an equal and opposite reaction.
   D. Every effect has a cause.

8. Substances that submit a great deal of resistance to the passage of electricity through it is called:

   A. Fuse.
   B. Insulator.
   C. Conductor.
   D. Transformer.

9. Why is it beneficial to enter a smoke filled room crawling low to the floor?

   A. Because smoke will be radiated through the floor.
   B. Because smoke is lighter than air.
   C. Because smoke may become flame.
   D. Because smoke will be compressed in a room.

10. A tank that weighs 500 pounds in the air appears to weigh 300 pounds when at a depth of 10 feet under water. If the tank is lowered to a depth of 20 feet, what will its apparent weight be?

    A. 100 pounds.
    B. 200 pounds.
    C. 300 pounds.
    D. 400 pounds

11. Of the following which is least likely to explode:

    A. Shut can of water over a high heat source.
    B. A can full of gasoline.
    C. A can of gasoline that is half empty.
    D. Gasoline can that has just been emptied.

12. The main reason to insulate pipes in heating:

    A. Is to prevent the loss of heat.
    B. Is to prevent pipes from rusting.
    C. Is to prevent fires.
    D. Is to prevent injury to people.

13. Which of the following is the best heat conductor:

    A. Copper.
    B. Glass.
    C. Steel.
    D. Water.

14. Which of the following represents the freezing point on a Fahrenheit thermometer:

    A. Zero degrees.
    B. 12 degrees.
    C. 30 degrees.
    D. 32 degrees.

15. An electron is:

    A. An electronic nucleus.
    B. A hydrogen atom.
    C. A static charge.
    D. A particle of negative electricity.

16. The exhaust gasses of an automobile consist of which of the following gases:

    A. Steam.
    B. Carbon, carbon dioxide, and carbon monoxide.
    C. Nitrogen.
    D. All of the above.

17. Electrical pressure is expressed in terms of:

    A. Ampere.
    B. Pressure gauge.
    C. Volt.
    D. Ohm.

18. The chain and the sprocket wheels of a bicycle increase the:

    A. Power of the rider.
    B. Force applied to the rear wheel.
    C. Force applied to the road.
    D. Speed of the rear wheel.

19. Of the following, what happens to water when it freezes:

    A. It contracts.
    B. It changes chemically.
    C. It expands.
    D. It increases substantially.

20. What is the term used to define the attraction that exist between the earth and the bodies on or near it:

    A. Force.
    B. Leverage.
    C. Vibration.
    D. Gravity.

21. Water will conduct electrical current better when:

    A. Its mineral content is at a high level.
    B. Its organic content is at a high level.
    C. Its mineral content is at a low level.
    D. It is completely pure.

22. Which of the following represents the weight of water:

    A. 12.50 pounds.
    B. 8.33 pounds.
    C. 5.50 pounds.
    D. 4.33 pounds.

23. Assume that you have two containers of equal size and shape. They are filled with equal amounts of water. A block of ice is added to one container and the same quantity of ice, chopped into cubes is added to the other container. What will happen to the water in the container with the cubes as compared to the container with the block?

    A. It will cool to the same temperature, faster.
    B. It will cool to the same temperature, slower.
    C. It will cool to a lower temperature, faster.
    D. It will cool to a lower temperature, slower.

24. Of the following, which is the worst substance to have contact a rope used for firefighting uses?

    A. Water.
    B. Oil.
    C. Gasoline.
    D. Acid.

25. Which science does hydraulics pertain to?

    A. Water/liquids under pressure.
    B. Water/liquids in motion.
    C. Gravity of water.
    D. Water for firefighting.

26. What is the substance that is usually used for electrical conductors?

    A. Rubber.
    B. Glass.
    C. Porcelain.
    D. Copper.

27. What is the term used to express the power of attraction of one chemical for another?

    A. Affinity.
    B. Oxidation.
    C. Gravity.
    D. Ventilation.

28. Materials that are incapable of combustion are:

    A. Flammable.
    B. Fireproof.
    C. Fire hard.
    D. Incinerated.

29. What would be the most likely cause of ignition to a pile of rags and trash laying in the corner of a room?

    A. Arson.
    B. Arching wires.
    C. Cigarette butt.
    D. Spontaneous combustion.

30. Petroleum is:

    A. Used only as a lubricant.
    B. Made from gasoline.
    C. Lighter than water.
    D. Used only as fuel.

31. Oils value as a lubricant depends upon its:

    A. Mixture.
    B. Origin.
    C. Viscosity.
    D. Color.

32. Which of the following would represent the greatest threat to an explosion:

    A. Square tanks.
    B. Round tanks.
    C. Full tanks.
    D. Almost empty tanks.

33. What unit of measurement does a millimeter represent?

    A. Volume.
    B. Weight.
    C. Viscosity.
    D. Length.

34. One millimeter equals what portion of a meter?

    A. One-thousandth of a meter.
    B. One-tenth of a meter.
    C. One half of a meter.
    D. One meter.

35. What does the electrical term "arching" refer to?

    A. An electrical current flowing along a conductor bent into a curve.
    B. An electrical current passing across a gap between two conductors.
    C. The blowing out of an electrical fuse.
    D. The blowing out of a circuit breaker.

36. When a material is referred to as combustible, it most likely is:

    A. Fireproof.
    B. Flammable.
    C. Crammed.
    D. Disintegrated.

37. Fire hose should not be cleaned with the use of gasoline because:

    A. It would cause it to rot.
    B. It would weaken the couplings.
    C. It would stain it.
    D. It would shrink it.

38. The smallest unit that a substance may be divided into without destroying the substance is:

    A. Atom.
    B. Particle.
    C. Nucleus.
    D. Molecule.

39. Which of the following oil soaked rags is most susceptible to spontaneous ignition:

    A. Motor oil.
    B. Castor oil.
    C. Linseed oil.
    D. Fuel oil.

40. The occurrence of unusual percentages of carbon monoxide at smoldering fires would most likely be due to:

    A. Oxygen deficiency.
    B. Very low temperature.
    C. Very high temperature.
    D. Similarity of materials.

41. When oxygen concentration in the air drops below 16 per cent:

    A. No flames will continue to burn.
    B. Most flames will be extinguished.
    C. Most materials will continue to burn freely.
    D. Breathing will not be affected.

42. When radioactive materials are involved in fires, the radiation coming from the fire:

    A. Increases in rate and intensity.
    B. Decreases in rate and intensity.
    C. Increases in rate and decreases in intensity.
    D. Is not affected.

43. As a fire increases in temperature, the percentage of carbon monoxide in the gases liberated:

    A. Remains constant.
    B. Increases.
    C. Decreases.
    D. Decreases at a fast rate.

44. Geiger counters will react to:

    A. Both alpha and beta, but not gamma radiation.
    B. Both alpha and gamma, but not beta radiation.
    C. Both beta and gamma, but not alpha radiation.
    D. Alpha, beta, and gamma radiation.

45. Hot magnesium, reacting with water, liberates:

    A. Manganous chloride.
    B. Nitrous oxide.
    C. Hydrogen.
    D. Nitrogen.

46. Natural gas consist principally of:

    A. Propane.
    B. methane.
    C. ethane.
    D. Butane.

47. Of the following flame colors, which one would indicate the highest temperature:

    A. Orange-red.
    B. Light red.
    C. orange-yellow.
    D. Yellow-white.

48. The freezing point on a Fahrenheit thermometer is:

    A. Zero degrees.
    B. 10 degrees.
    C. 28 degrees.
    D. 32 degrees.

49. An electron is a/an:

    A. Particle of negative electricity.
    B. Static charge.
    C. Hydrogen atom.
    D. Electronic nucleus.

50. Any substance which offers a very high or very great resistance to the passage of electricity through it is called a/an:

    A. Fuse.
    B. Conductor.
    C. Insulator.
    D. Closed circuit.

ANSWER SHEET

CHEMISTRY AND PHYSICS DIAGNOSTIC TEST

1. A	26. D
2. C	27. A
3. C	28. B
4. B	29. D
5. B	30. C
6. D	31. C
7. C	32. D
8. B	33. D
9. B	34. B
10. C	35. B
11. B	36. B
12. A	37. A
13. A	38. D
14. D	39. C
15. D	40. A
16. D	41. B
17. C	42. D
18. D	43. B
19. C	44. C
20. D	45. A
21. A	46. B
22. B	47. D
23. A	48. D
24. D	49. A
25. B	50. C

SUMMARY

Questions relating to Chemistry and physics will be found on general aptitude entrance test and job related entrance test. It is necessary for you to have a good knowledge of the terms discussed in this chapter and why the terms are important to Firefighters.

ASSIGNMENT:

1. Information Sheet #12: "Fire Prevention and Building Construction", on pages 366 - 370.

CHAPTER 10

MECHANICAL COMPREHENSION

MECHANICAL INFORMATION

INTRODUCTION

The job of firefighting consists of many tasks that involve mechanical performance. Therefore many entrance level examinations will have a section that will require the candidates to demonstrate their knowledge and abilities relating to mechanical comprehension. In this chapter we will discuss mechanical theories and concepts. topics identified will be basic electronics, the basic elements of life identified as the atom and its components, hand tools and equipment, basic vehicle mechanics, pulleys, belts and wheels.

GENERAL INFORMATION

BASIC THEORIES AND CONCEPTS:

When two objects of the same size, but of different weights, are dropped from a given height, both objects will reach the ground at the same time because the effect of gravity is the same on both objects.

The air in the highest portion of a room is normally higher in temperature than the air near the floor because warmer air is lighter that cold air.

When using an auger bit to drill a hole through a piece of wood, it is a good idea to clamp a piece of scrap wood on the underside of the piece that is being worked on so as to prevent it from splintering.

An electric fuse is used to protect against overload.

Electrical conductors are generally made of copper.

An electron is a particle of negative electricity.

Any substance which offers a very high or very great resistance to the passage of electric current through it is called an insulator.

Water is a better conductor of electricity when its mineral content is high.

ELECTRICAL SERVICE ENTRANCE CONSIST OF:
1. Service drop, service head, and insulators.
2. Conduit or cable.
3. Weather proof meter and socket.
4. Entrance switch with circuit breakers.
5. Ground wire.

Electrical VOLTAGE is like PSI (pressure).

Electrical AMPERES is like GPM (rate of flow).

Electrical OHMS is like FL (friction loss, resistance).

ELECTRICAL AMPERES is the most hazardous component of a high voltage electrical circuit.

OHMS LAW:
1. Amperes = volts divided by ohms.
2. Volts = amperes multiplied by ohms.
3. Ohms = volts divided by amperes.

OVERLOADING an electrical circuit will cause overheating.

Maximum continuous current an individual may safely be subjected to is FIVE (5) MILLIAMPERES.

TRANSFORMERS reduce voltage.

There is an INCREASE of life hazard with the magnitude of voltage.

PROTONS are positive charged and NEUTRONS have no charge.

LOOSELY BOUND ELECTRONS are good insulators. (there are no perfect insulators).

In TRANSFORMER FIRES, let fire burn itself out.

IONIZED AIR is the most conductive of electrical charges.

In electrical distribution systems, PRIMARIES are the high voltage lines and SECONDARIES are the low voltage side of the system.

The voltage side of the transformer to home or place of business is equal to 600 VOLTS or less.

Firefighters MAY CUT power lines up to 8000 VOLTS.

Firefighters must be familiar with the hazards of electrical currents. The LOWEST that could be lethal to a well grounded person is 110 VOLTS.

There is little danger to firefighters directing hose streams on wires of LESS THAN 600 VOLTS to ground from any distance likely to be met under conditions of ordinary firefighting.

When cutting electrical service lines while the wires are overhead, the firefighter should, whenever possible, CUT THE LOOPS attached to the service head.

When an electrical wire has been pulled apart with NO CURRENT flowing, usually the break will have a CUP effect on one end and a CONE effect on the other end.

When an electrical wire has been cut with NO CURRENT flowing, usually it will have TWO WEDGE LIKE ENDS.

While an electrical wire IS CARRYING CURRENT during the break, it usually will have BOTH ENDS FUSED AND DISTORTED.

While fighting electrical fires, it must be remembered that water under high pressure becomes a GOOD CONDUCTOR.

RUBBER BOOTS should not be relied upon to supply the necessary resistance to ground, to prevent completion of electrical circuit through the body. (boots may contain carbon black).

If current doubles, heat will be FOUR TIMES AS GREAT.
ELECTRICAL GROUNDING: a connection between a conductive body and the earth that deletes the difference in potential between the article and ground.

There is a definite relationship between the LENGTH OF TIME (exposure) to electrical shock and the effects of the shock.

ELECTRIC TRANSFORMER: apparatus that converts and
delivers electricity received at a certain voltage to electricity of a different voltage.

A shaft, within a building, is best defined as an enclosed space connecting a series of two or more openings in successive floors.

The property of matter by reason of which a body tends to resume its original shape when changed by an external force is termed as elasticity.

The force which holds together bodies of the same kind is termed as adhesion.

The term which is applied to the distance from the center of a circle to its boundary is known as the radius.

Automatic sprinkler systems are usually set in operation by heat.

Lubricant prevents rubbing surfaces from becoming hot because it forms a smooth film between the two surfaces, which prevents them from coming into contact.

A "spot weld" is when parts are held together in different places to hold them in place.

Pavement may crack during hot weather because heat expands.

Rust on tools and equipment is the result of oxidation.

TOOLS AND EQUIPMENT:

All purpose tools may be used for a great variety of purposes.

Special tools are suitable only for a particular purpose.

AXE HANDLES are usually made of wood rather than steel because the wood tends to cushion the impact.

PICKHEAD AXE: a hand axe with a blade and a pick, also called FIREMAN'S AXE.

FLATHEAD AXE: is a single bitted axe with a flattened head.

PIKE POLE: is a wooden or fiberglass pole that has a metal point with hook. (usually used to pull down plaster from ceilings).

For opening of ceilings from below, use PIKE POLE or PLASTER HOOKS.

A CARBIDE-TIPPED CHAIN SAW is designed to cut:
1. Clay brick.
2. Asphalt.
3. Composition roofing material.

When using a SAWZALL the SHOE should be held FLAT against the material to be cut.

HOSE CLAMPS are used for shutting down the water from charged lines. The hose clamp is placed a minimum of 25 feet behind the apparatus and approximately 6 feet behind the coupling on the hydrant side.

Besides its use for opening doors that swing inward, the DETROIT DOOR OPENER is effective for the use as an emergency hose clamp.

FOUR-WAY VALVE: used on a hydrant primarily to permit the Fire Engineer to change from direct hydrant stream to pump stream without shutting of the water.

HALYARD: the rope used to extend an extension ladder.

MANILA ROPE is the standard for Fire Department rope.

COTTON FIBER ROPE: tensile strength is less than nylon fiber, sisal fiber, and manila fiber.

If there is a knot in a rope the rope will be WEAKENED at that point.

A temporary method of securing an object with a rope so that it can be easily undone is called a HITCH.

EYE SPLICE loop spliced at end of rope.

BOWLINE KNOT: will not slip or tighten under tension and it is easily untied. multipurpose knot.

The BEST method for holding the LIFE NET is with both palms facing the firefighter and thumbs alongside the first finger.

SCBA: Self-Contained Breathing Apparatus have no maximum time limit for changing air bottles as long as the pressure remains sufficient.

Air pressure in demand type breathing apparatus is 2000 PSI TO 2300 PSI.

Standard compressed air cylinders (steel) are regulated by INTERSTATE COMMERCE COMMISSION (ICC).

Firefighters MAY NOT BE ABLE to ware breathing apparatus while responding to an incident, because the seat belts or safety tailboard straps may not work while wearing breathing apparatus.

Self Contained Underwater Breathing Apparatus (SCUBA) contain a mixture of OXYGEN AND NITROGEN.

Standard compressed air cylinder for breathing apparatus is rated for 30 MINUTES.

SEQUENCE FOR USING STRAPS ON BREATHING APPARATUS FACE MASK:
    1. Neck.   2. Side.   3. Top.

Portable Oxyacetylene torch (cutting), Oxygen is set at 35 PSI.

PITOMETER: used to determine the flow capacity of fire hydrants.

PYROMETER: used for measuring high temperatures.

DECIBELS: used in measuring amounts of sound.

GALVANOMETER: used for measuring current-strength or potential difference.

GALVANOMETRY: is the science, art or process of measuring electric currents.

GALVANOMETER: is activated by electricity.

The HURST POWER TOOL uses a GASOLINE ENGINE.

MECHANICAL INFORMATION:

WORN brake lining will cause squeaky brakes.

Unequal adjustment of brakes creates danger of SKIDDING.

BRAKE FADING on fire apparatus is caused by the brake drums and lining getting hot.

Brake lining on fire apparatus will SQUEAL if foreign material gets embedded in it.

EXCESSIVE HEAT is the cause of brake fade, which results in poor braking due to the rapid wearing away of the lining.

ORDINARY BRAKE FLUID is not used in the disk air brake system because it will break down too fast.

At 80 degrees F batteries are fully charged with a specific gravity reading of 1.280 on the hydrometer.

BATTERY ELECTROLYTE is sulfuric acid and water.
A discharged battery has a LOWER specific gravity.

Gases released from storage batteries are HYDROGEN and OXYGEN.

Large amounts of water added to battery would indicate DEFECTIVE voltage regulator.

CURRENT FLOW in storage battery is drawn from positive to negative.

ELECTROLYTE is the liquid in a storage battery.

NEUTRALIZE battery acid with BICARBONATE OF SODA.

If a lead acid battery is discharging, the electrolyte will become WEAKER.

In a fully charged battery, the ELECTROLYTE is
1 1/4 TIMES AS HEAVY as pure water.

SULFURIC ACID is the acid used in lead acid storage batteries.

While replacing a battery, if you have a question as to the correct pole to ground, connect the battery and check the AMMETER.

A high charging rate is indicated by high battery water CONSUMPTION.

A lead acid storage battery with six cells connected in series will produce 12 VOLTS.

Keeping the plates covered in a battery is the MOST IMPORTANT thing that you can do to prolong the life of the battery.

The VOLTAGE in batteries can be increased by increasing the number of CELLS.

Operating fire apparatus electrical accessories with the engine OFF is a severe DRAIN on the battery.

Current from a storage battery is drawn from POSITIVE to NEGATIVE.

While a storage battery is recharged it liberates HYDROGEN.

When removing a battery, REMOVE THE GROUNDED TERMINAL FIRST.

High battery water CONSUMPTION usually indicates a high charging rate.

A HYDROMETER has a scale calibrated to read SPECIFIC GRAVITY.

Electrolyte of FULLY CHARGED BATTERY weighs 1 1/4 times as much as pure water.

Fully charged battery will LOSE 10% of its power at 0 degrees F.

OVERCHARGING is the most common cause of battery damage.

Per-cent of BATTERY CHARGE to specific gravity:
    1.265 to 1.280 = 100%
        1.225 = 75%
        1.200 = 55%
        1.190 = 50%
        1.155 = 25%

FUSES and CIRCUIT BREAKERS protect against amperage overload.

The PRIMARY function of the CURRENT REGULATOR, as distinct from the voltage regulator, is to prevent the generator from charging beyond maximum-rated output.

The AMMETER on a fire apparatus shows the flow of the electric current to and from the storage battery.

If AMMETER needle is FLUCTUATING rapidly, the most likely cause is a faulty REGULATOR.

The PRIMARY function of the cutout relay in a GENERATOR is to keep the battery from discharging while the engine is idling.

The electrical system of an ALTERNATOR is prevented from reversing its flow of electricity from the battery by the use of a RECTIFIER.

PRIMARY function of voltage regulator is to keep battery from overcharging.

IGNITION SWITCH completes connection between battery and coil. (primary lead).

IGNITION COIL increases amperage and decreases voltage.

MAGNETO generates electric current.

CAPACITOR is part of the distributor.

CONDENSER is used in the ignition system.

DWELL ANGLE is the number of degrees a cam rotates while ignition points are closed.

PISTON: the part in a cylinder that moves up and down.

COMPRESSION GAUGE is used to check the cylinder pressure.

Cam follower is the VALVE LIFTER.

COIL: the device used to step-up low voltage to high voltage needed at the spark plugs.

DISTRIBUTOR: delivers the spark directly to the spark plugs.

CARBURETOR: is where gasoline vapor and air are mixed to the correct proportion.

Oil DILUTION is prevented in the crankcase by ventilation.

The OIL GAUGE on the dashboard indicates oil pressure within the engine.

RADIATOR pressure cap is used to raise the boiling point of the water in the radiator.
As far as VISCOSITY, the LOWER S.A.E. rating is, the easier the oil will flow at lower temperatures.

THERMOSTAT: usually operates with a bimetallic strip.

The primary purpose of a THERMOSTAT is to shut off the water circulation between the radiator and the engine, when the engine is cold.

SPEEDOMETER and TACHOMETER both use permanent magnets.

Excessive wear on the outside edge of a front tire is likely to be caused by too much CAMBER.

POSITIVE CASTER will be caused by tipping the top of the kingpin backward from the vertical position.

TOO LITTLE free movement in the clutch pedal, clutch may not engage.

TOO MUCH free movement in the clutch pedal will result in failure of the clutch to disengage properly.

Riding the clutch INCREASE the possibility of damaging the throw out bearings.

Thermostat has a BI-METALLIC strip.

Frequent and regular washing, waxing, and polishing will LENGTHEN the life of the painted finish and bright metal trim. (warm or cold water).

IDEAL engine operating temperature is just below the boiling point of the liquid being used.

RING GEAR is located in the standard differential.

A fire apparatus can go FASTER forward than in reverse because the gear ration is almost as high as low gear.

The TRANSMISSION is responsible for allowing the change in gear ratio between the engine and the rear wheels of fire apparatus.

Higher COMPRESSION ratio will give greater power at all speeds.

CAM FOLLOWER is a valve lifter. CAM SHAFT operates the valve push rods.

Rotative force developed by engine is TORQUE.

The MAIN PURPOSE of a fire department pumper is to provide adequate pressure for fire streams.

To prevent engine damage while driving downhill, the speeds should be not more than 200 to 300 RPM above rated speed.

Oil gage needle FLUTTERING may indicate low oil.

Oil filters REMOVE sludge.

CHOKE: To restrict flow of a fluid.

EMULSIFIED OIL is caused by water vapor from burning fuel.

Racing gasoline engines prior to turning off ignition will DILUTE crankcase oil.

Water formed by burning fuel causes CORROSION of internal engine components.

Main effect of DETERGENTS added to lubricants is to hold contaminants in suspension.

Perfect gasoline to air mixture = 15 PARTS AIR to 1 PART GASOLINE, at idle = 11 parts air to 1 part gasoline.

The device which gasoline vapor and air are mixed in proper proportion is known as the CARBURETOR.

CHOKE VALVE is for rich fuel-air mixture. (shuts off air intake).

Engine missing at high speeds is caused by partly stopped-up FUEL LINE.

AIR CLEANER will act as a flame arrester in case of engine backfire through the carburetor.

The GREATEST SOURCE of engine oil contamination is normally from unburned fuel.

In an internal combustion engine the DETONATION is the result of the secondary ignition of the fuel charge after the regular spark occurs.

The HIGHER the octane rating of a fuel, the higher its ability to resist detonation.

Internal combustion engines having CONSTANT VOLUME combustion are usually those utilizing gasoline and air.

CHOKE DAMP: mixture of gases causing choking.

Apparatus engines should be checked for proper oil level with the ENGINE OFF, and approximately 30 minutes after shutting off.

Engine cutting is usually CARBURETOR problems, engine cutting at high speeds is caused by partly stopped-up gas line.

TOO RICH fuel causes:
1. Loping-sluggish engine.
2. Irregular running engine.
3. Engine overheating.
4. Black smoke.
5. Dirty air cleaner.

TOO LEAN of carburetor fuel mixture causes:
1. Motor popping back into manifold and carburetor.
2. Engine too slow or stops when accelerated suddenly.
3. Noticeable loss of power.

The PURPOSE of ventilating the crankcase is to prevent oil dilution.

When an engine will not start even though the starter turns over, the trouble is most likely the FUEL SUPPLY.

Power developed by GASOLINE ENGINES decreases 3 1/2 % per 1000 foot altitude above sea level.

Power developed by DIESEL ENGINES decreases 3% per 1000 foot altitude above sea level. (normally aspirated diesel engines).

TURBOCHARGED DIESEL ENGINES do not have a power loss until altitudes in excess of 4000 feet and then only 2% per 1000 foot altitude above 4000 foot altitude level.

The PRINCIPAL advantage of obtaining a higher compression ratio in an fire apparatus engine is the engine will have greater power at all speeds.

Perfect GASOLINE TO AIR MIXTURE is 15 parts air to 1 part gasoline, at idle it is 11 parts air to 1 part gasoline.

Regulate speed and power in diesel engines by AMOUNT of fuel in the cylinders.

The quickest and safest way to find a spark plug not firing is to SHORT the plug to the engine with a wooden handled screw driver.

A partly stopped up fuel line may be indicated by the engine MISSING at high speeds.

VAPOR LOCK can be treated by moving the fuel line away from the exhaust manifold.

Ignition quality of diesel fuel is measured by OCTANE.

DIESEL engine detonation is usually EARLY fuel injection timing, when related to the injection timing.

COMPRESSION IGNITION is used in diesel engines.

To change engine speed or power in diesel engines you must VARY THE AMOUNT OF FUEL, the air entering always remains the same.

In most 4 STROKE engines 2 revolutions of the flywheel are completed with each cylinder cycle.

In 6 CYLINDER 4 CYCLE gas engines the breaker cam and distributor rotor will rotate at 1/2 the crankshaft speed.

In a gas engine that one explosion takes place in each cylinder once every two revolutions of a shaft is called a 4 CYCLE ENGINE.

LUBRICANTS prevent rubbing surfaces from becoming hot because of the smooth film that it creates between the surfaces, which prevent them from coming into contact.

If the OIL PRESSURE GAGE drops to 0 the FIRST thing that an apparatus driver should do is STOP the engine to prevent damage to the bearings.

To PREVENT oil dilution fire apparatus crankcases are ventilated.

The HIGHER the SAE number of engine oil = the higher the viscosity.

DETERGENTS in lubricants HOLD CONTAMINANTS IN SUSPENSION.

Engine overheating is usually 1 MAJOR PROBLEM rather than several minor problems.

Excessive wear of the outer edge of the tread of the front tires is most likely the result of too much CAMBER.

TIRE WEAR:
1. Under inflated = both edges wear
2. Overinflated = center of tread wear
3. Out of line = wear on one edge
4. Improper balance = flat spots, cupping
5. Heavy braking = flat spots, abrasions
6. Toe in or out = feathering
7. Fast cornering = feathering
8. Camber = tire wear on outside

Tire UNBALANCE causes shimmy.

CAMBER is the inward inclination of the front wheels at the bottom. Wear is on the outer edge. (refers to wheel adjustment).

The MOST important type of friction in the control of an apparatus is between the road surface and the tires.

A bumpy road REDUCES available friction between the road surface and the tires.

CAMBER refers to the adjustment of the wheels of the vehicle.

In wheel alignment, the amount of TOE-IN is adjustable by changing the length of the tie rod.

The TILLERMAN on a fire apparatus is for controlling the rear wheels.

Driving habits of operators are the major factors for obtaining maximum FUEL ECONOMY from vehicles.

HIGHEST MANIFOLD VACUUM when operating an engine under its own power is obtained oat INTERMEDIATE SPEED, with the throttle partially opened.

On HEAVY APPARATUS the greatest driving torque reaction is transmitted through the drive train and is absorbed by the REAR SPRINGS.

IGNITION COIL: takes the current from the battery and generator and delivers it to the spark plugs with INCREASED AMPERAGE and DECREASED VOLTAGE.

ROUND WIRE FEELER GAUGE: is the most accurate tool to use to set the proper gap on a spark plug.

Measure the COMPRESSION in an internal combustion engine at the SPARK PLUG HOLE.

PRE-IGNITION, with ignition switch off is most likely caused by an overheated engine.

MAXIMUM TORQUE is normally obtained from an engine at speeds BELOW the engine speeds at which maximum horsepower is produced.

ONE HORSEPOWER is the power required to lift 550 LBS 1 foot in 1 second.(33,000 LBS 1' in 1 minute)

CAM SHAFT: operates the valve push rods.

The RING GEAR is located in the STANDARD DIFFERENTIAL.

A 4-stroke cycle gas engine with a crankcase gear of 24 teeth will have a TIMING GEAR on the camshaft of 48 TEETH.

SPUR GEARS in conventional transmissions have teeth that appear to run straight across the wheel.

The BEVEL GEARS most common use is to transmit power from two shafts whose axes intersect at right angles.

Fire Department storage batteries fluid = DILUTED SULFURIC ACID.

AMMETER: measures the amperage of current. RECTIFIER works with the alternator.

ELECTROLYTE: used in a wet cell battery, is a chemical compound which will decompose when an electric current is passed through it.

Generator, transformer and voltage regulator are used to GENERATE or CONTROL electrical currents.

OVERCHARGING THE BATTERY is the most frequent cause of battery failure in the Fire Service.

CURRENT REGULATOR protects the generator. VOLTAGE REGULATOR protects the battery.

Ideal operating engine temperature for liquid cooled engines is JUST BELOW THE BOILING POINT of the liquid being used.

AMBIENT TEMPERATURE of vehicle engine = temperature inside the engine compartment.

MECHANICAL PRINCIPLES:

PULLEYS:

1. The more pulleys the easier it is to pull or lift an object.
2. The more pulleys involved the greater distance you must pull, but it is still easier to lift an object.
3. The thinner a windlass, the easier it is to turn.
4. In two different sets of pulleys, if the wheels are connected by a shaft, and the two wheels on one pulley are the same as the other two that they are connected to, then they both turn at the same speed.

BELTS:

1. Always determine in which direction one of the wheels in a diagram is turning as the belt will be going in the same direction. Also you can determine the direction of the belt and the wheel direction will be the same.
2. Wheels under a belt that is not twisted all turn in the same direction. Those on the outside of the same belt would turn the opposite direction of those on the inside.

WHEELS:

1. If wheels are a different size on the same vehicle, then the smaller wheel will turn faster.
2. When wheels of different sizes are joined together by belts, the smallest wheel turns fastest, the largest wheel slowest.
3. When two gears of different sizes are locked together, the smaller gear turns faster than the shaft connected to the larger gear.

TURNING/DIRECTIONS:

1. The faster an object whirls around, the more it will pull from the center of rotation.
2. If a car, or tractor or objects are turning, then the inside wheels or objects will turn less distance and more slowly than the outside ones.
3. When a car skids, its speed increases momentarily to the outside when turning.

CENTER OF GRAVITY: (referring to the point which weight is evenly distributed)

1. A solid object with a space drilled out will rest on the section that is solid.
2. The higher a vehicle is packed with materials the easier it will turn over when on an incline.

VOLUMES AND AREAS OF SOLID OBJECTS:

1. If several solid figures have the same width and height but different shapes then their weight and volumes are different. The lowest weight or least volume is a solid of triangular shape. Then comes a cylindrical solid (circular in shape) and then a cube (square shape)
2. Objects (cars) placed or parked parallel to each other and perpendicular to the side occupy less space.

SUMMARY

Once again we have identified information regarding mechanical aptitude that you should be familiar with. If necessary additional information regarding these topics can be obtained at your school or community Library.

ASSIGNMENT:

1. Information Sheet: "Extinguishing Systems", on pages 371 - 377.

CHAPTER 11

MECHANICAL APTITUDE

MECHANICAL APTITUDE

INTRODUCTION

In this chapter there is a diagnostic test that deals with mechanical aptitude. When answering mechanical type questions, base your answer on the information provided with each question, and nothing more. The correct answer for the question will only relate to the stated problem. Even if you consider that the information presented is not adequate or up to date, base your answer on the information provided.

Read each question carefully, paying particular attention to questions with drawings. You will need to carefully examine the questions, the diagrams, and the information that goes with the diagrams.

When solving problems involving illustrations, stay composed. Even if the illustration looks unfamiliar, difficult, or the question appears obscure. Carefully study the question in connection with the illustration. This will help you figure out the answer.

When encountering mechanical type questions: base your response on the information provided with each question, and nothing more! The correct answer for the question will only relate to the stated problem. Even if you consider that the information presented is not accurate or up to date, base your answer on it and it only.

Remember when taking an examination and you encounter Science, Chemistry, or Physics type questions: don't be afraid if your knowledge is not very broad. General knowledge of the basic concepts will be adequate. If you find it necessary to improve your knowledge in these areas, you might consider taking a class in basic chemistry and/or physics.

MECHANICAL APPTITUDE DIAGNOSTIC TEST

QUESTIONS: PULLEYS, GEARS, AND LEVERS

1. Of the two block & tackles below, which one will lift a load more easily?

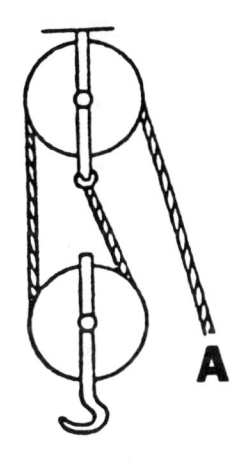

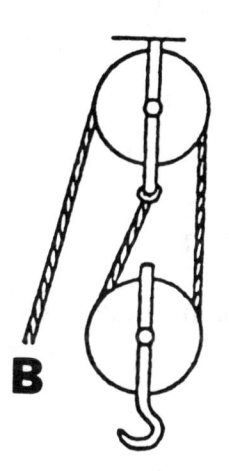

    A. Drawing #1   B. Drawing #2   C. No difference.

2. Cylinder #1 and cylinder #2 turn in opposite directions at the same time. If #1 turns in the direction of the arrow, in what direction will the pulley hook travel?

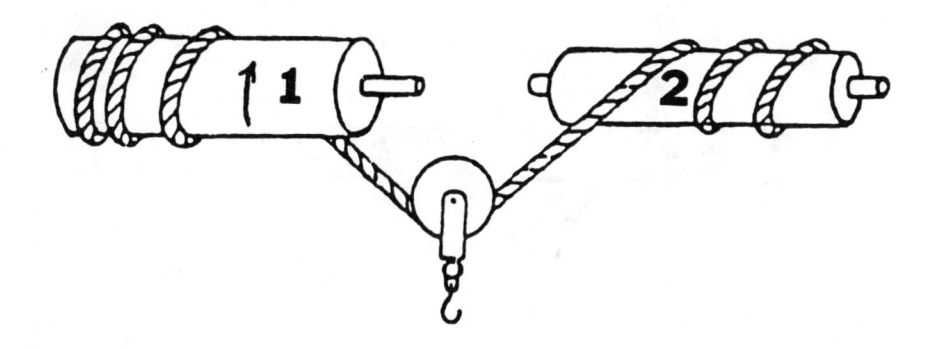

    A. In an upward direction.
    B. In a downward direction.
    C. It will not travel up or down.

3. Of the two drawings below, which winch will be able to lift a heavier load?

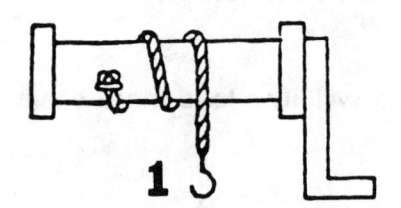

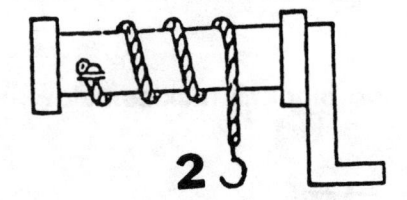

A. Drawing #1.
B. Drawing #2.
C. They will lift equal loads.

4. In the drawing below, if the cylinder is rotated in the direction of the arrow, in what direction will the pulley hook go?

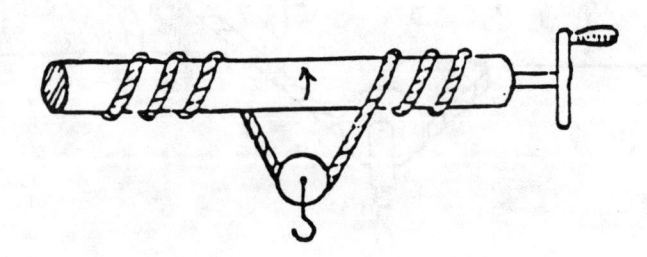

A. Downward.
B. Upward.
C. Neither direction.

320

5. If the large cylinder turns in the direction of the arrow, in what direction will the pulley hook go?

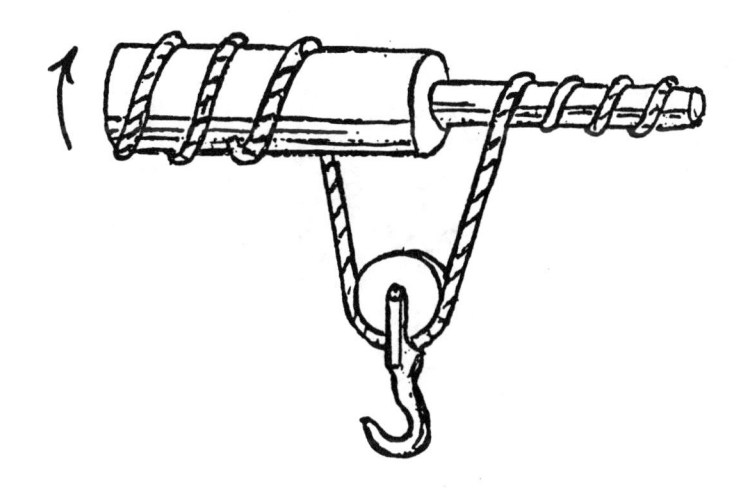

   A. Upward.
   B. Downward.
   C. Neither direction.

6. In which direction must the larger gear rotate in order to make the "C" rotate?

   A. Counterclockwise.
   B. Clockwise.
   C. Either counterclockwise or clockwise.

7. If gear "X" is rotated in the direction of the arrow, which direction will gear "Y" rotate?

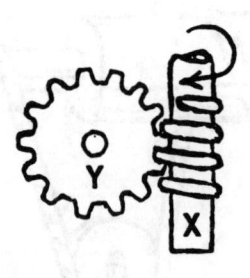

A. Clockwise.
B. Counterclockwise.
C. Either clockwise or counterclockwise.

8. In the drawing below, which direction can the sprocket rotate?

A. Counterclockwise.
B. Clockwise.
C. In both directions.

9. Which direction must the inside gear be rotated in order to rotate the outside gear?

A. Counterclockwise.
B. Clockwise.
C. In either direction.

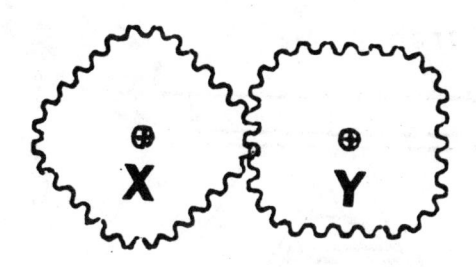

10. In the drawing below:

A. For each turn of gear "X", gear "Y" will make a complete rotation.
B. Only gear "X" can be the driving gear.
C. Wheel "Y" will move more slowly than Wheel "X".
D. None of the above.

11. In the drawing below:

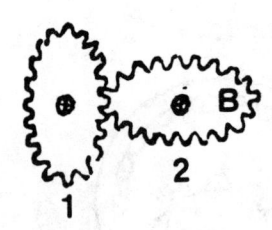

A. When gear #1 touches gear #2 at the "B" position, gear #2 will be moving slower than gear #1.
B. The gears will not turn.
C. Gear #2 can be the driving gear.
D. None of the above.

12. In the drawing below, when a force is exerted at the location of the #1, on the teeterboard drawing below, in which direction will end #2 go?

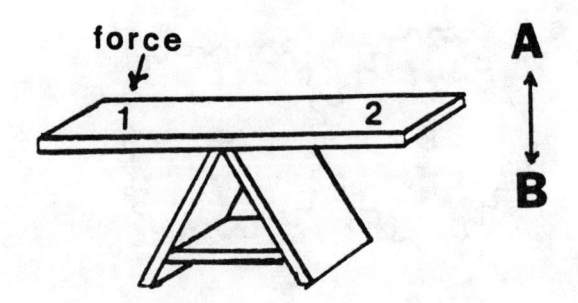

A. Up.
B. Down.
C. Neither direction.

13. In the drawings below, which load will be the easier to lift when exerting pressure at the location of the arrows?

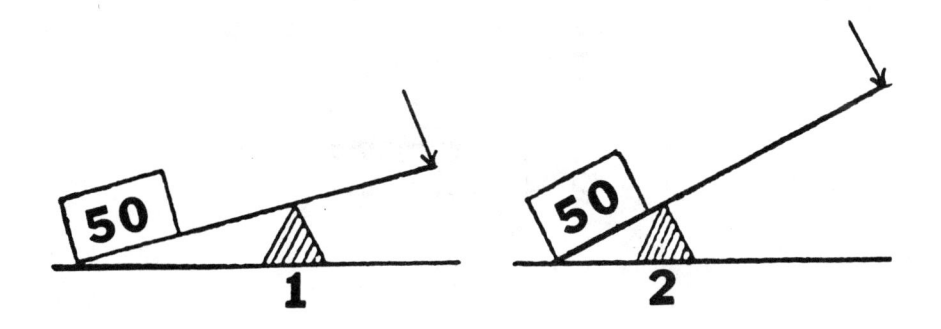

A. Drawing #1.
B. Drawing #2.
C. It will not make any difference.

14. In the drawing below, which block will be the most difficult to tip over?

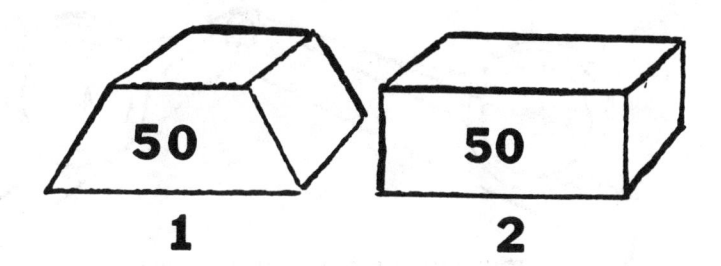

A. #1.
B. #2.
C. There would be no difference.

15. In the drawing below, which chain has the greatest stress on it?

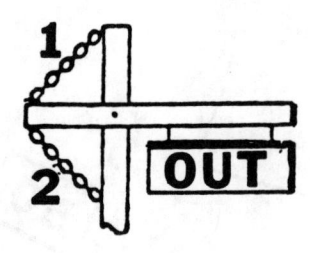

A. Chain #1.
B. Chain #2.
C. They have equal stress.

16. In the drawing below, if #1 and #2 are of equal size and weight, and the pulley at the top will turn freely, which one of the pulleys will tend to roll down the ramp?

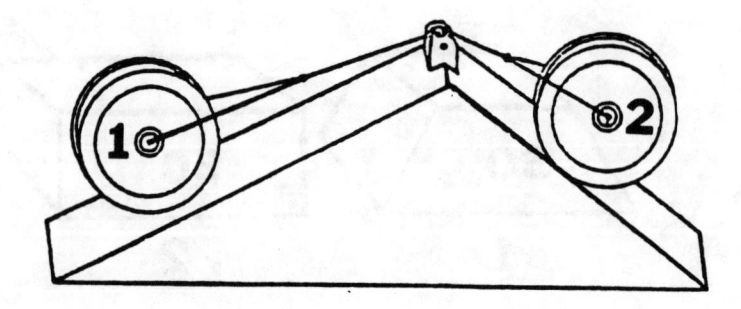

A. #1.
B. #2.
C. It will not make any difference.

17. In the drawing below, which ladder is the LEAST likely to tilt over?

A. Ladder #1.
B. Ladder #2.
C. It does not make any difference.

18. In the drawing below, which teeter would be the most stable?

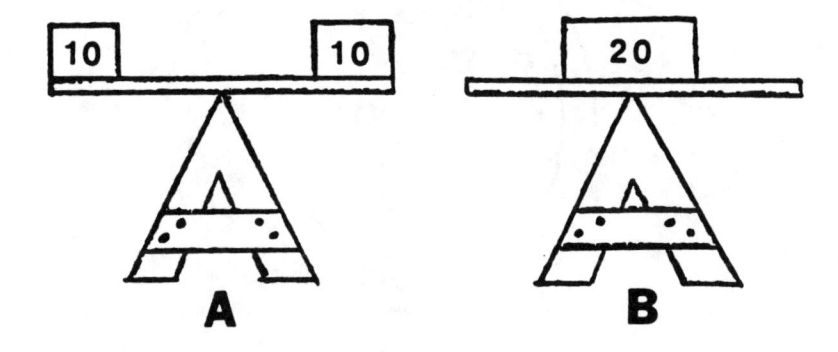

A. Teeter A.
B. Teeter B.
C. They are of equal stability.

19. In the drawing below, in order to SHORTEN the turnbuckle, in which direction should it be turned?

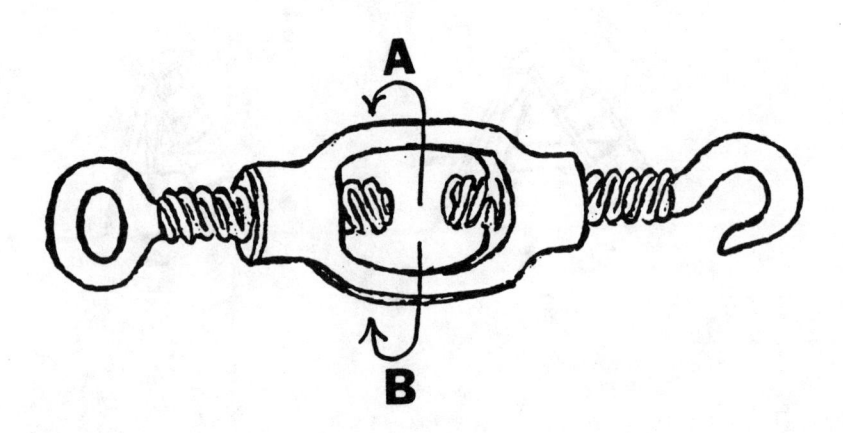

A. Direction A.
B. Direction B.
C. It would not make any difference.

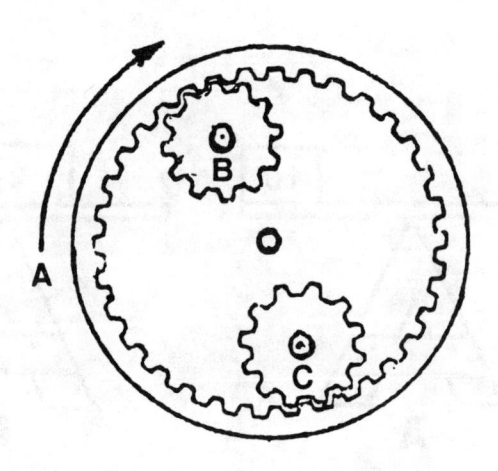

20. In the drawing below:

A. Gear A and gear B will turn in opposite directions.
B. Gear B and gear C will turn in opposite directions.
C. Answers A and B are both correct.
D. None of the above.

21. Which block and tackle will lift the load easier?

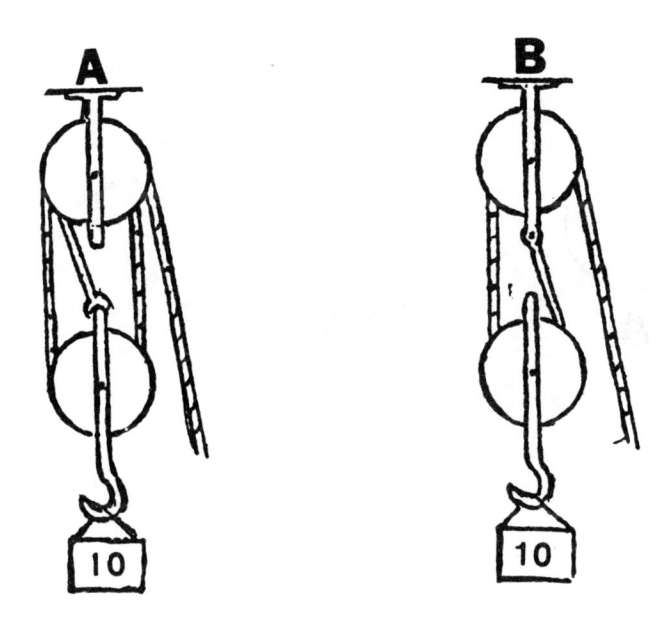

- A. Block and Tackle A.
- B. Block and Tackle B.
- C. They will lift at the same rate.

22. Which winch will raise the weight more easily?

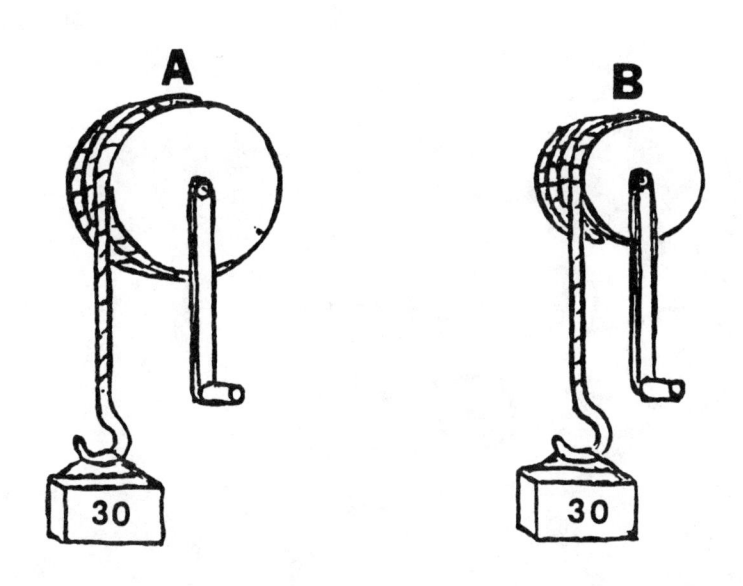

- A. Winch A.
- B. Winch B.
- C. They will both lift equally.

329

23. In the drawing below, if wheel A and wheel B make a complete revolution every second, which **NOTCH** will travel the longest distance?

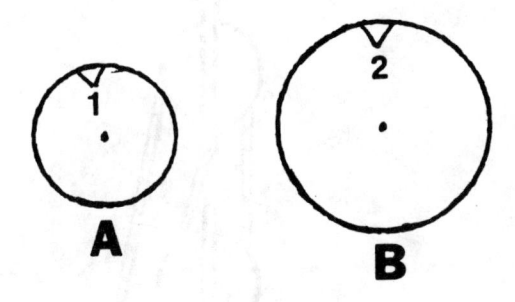

A. Notch #1.
B. Notch #2.
C. Both move at an equal rate of speed.

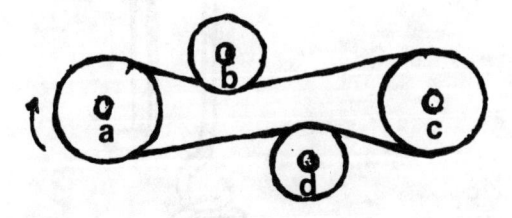

24. In the drawing below:

A. Wheels b and d turn in opposite directions.
B. This setup can run in either direction.
C. All four of the wheels will run at the same speed.
D. None of the above.

25. In the drawing below:

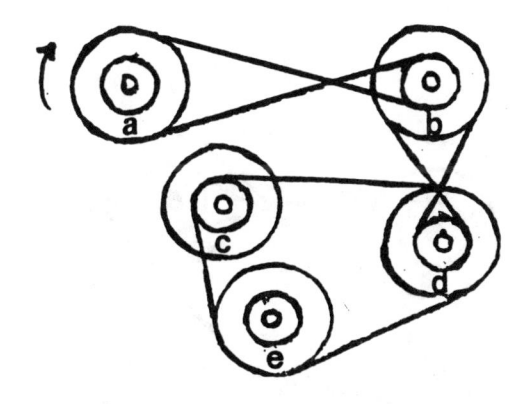

A. Wheels a and c turn in opposite directions.
B. Wheels a and d turn in opposite directions.
C. Wheel d turns faster than wheel a.
D. Wheel b turns faster than wheel e.

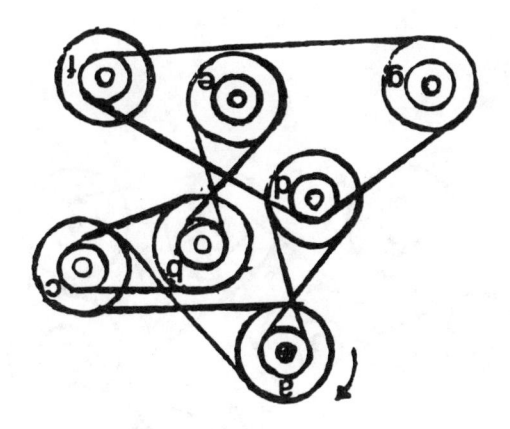

26. In the drawing below:

A. Wheels c and d turn in the same direction.
B. Wheels a and f turn at the same speed.
C. Wheel g turns faster than wheel b.
D. None of the above.

27. In the drawing below:

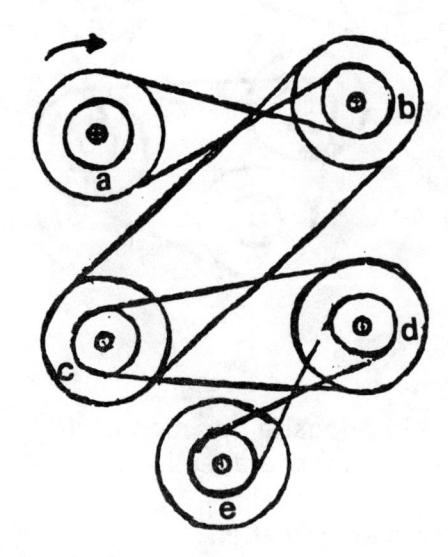

A. Wheels a and d turn in the same direction.
B. Wheel d turns faster than wheel b.
C. Wheel a turns faster than wheel c.
D. Wheels d and a turn at the same rate.

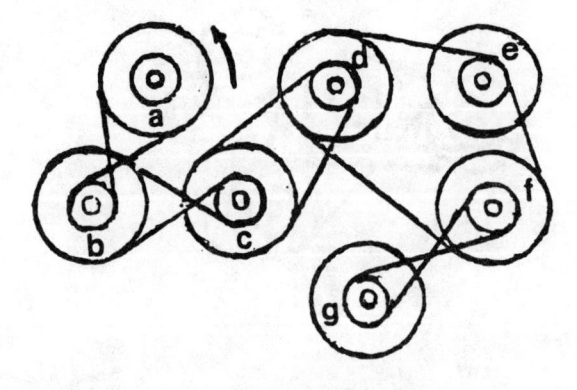

28. In the drawing below:

A. Wheels a and g turn in the same direction.
B. Wheels b and e turn in opposite directions.
C. Wheel c makes more revolutions than wheel g.
D. Wheel b makes more revolutions than wheel e.

29. In the drawing below, one complete turn of the drum crank will move the weight vertically upward a distance of?

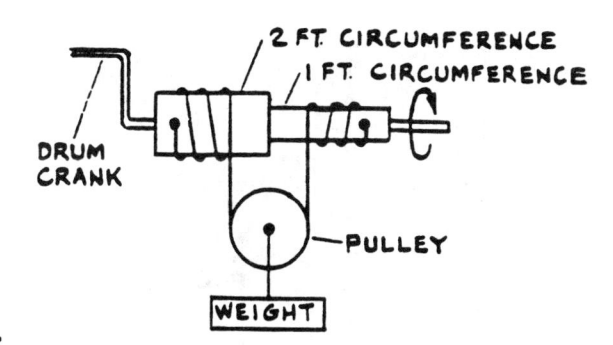

A. 3 feet.
B. 2 1/2 feet.
C. 2 feet.
D. 1 1/2 feet.

30. In the drawing below, the maximum weight (w) that can be lifted with a 75 pound pull as shown is?

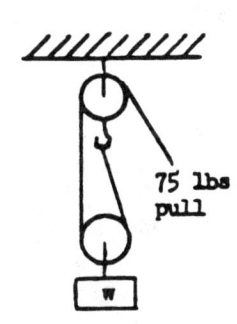

A. 50 pounds.
B. 75 pounds.
C. 100 pounds.
D. 150 pounds.

31. In the drawing below, which shaft will turn the slowest?

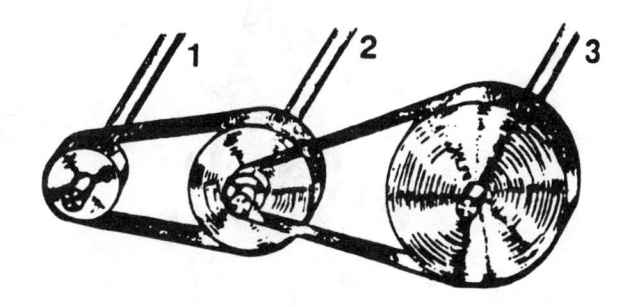

A. Shaft #1.
B. Shaft #2.
C. Shaft #3.
D. They will all rotate at the same speed.

32. In the drawing below, while lifting the load the required pull decreases:

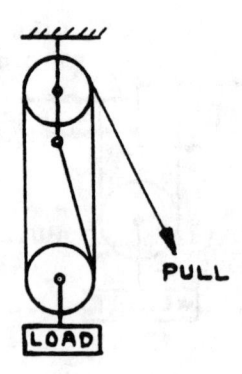

A. As the distance between the pulley wheels decreases.
B. When the rope size is increased.
C. When the number of pulley are increased.
D. If larger pulley wheels are used.

33. In the drawing below, when the top pulley rotates in the direction of the arrow, in what direction will the lower pulley rotate?

A. Direction A.
B. Direction B.
C. In either direction.

QUESTIONS 34, 35, AND REFER TO THE DRAWING BELOW. WHEELS "A" AND "B" HAVE THE SAME DIAMETER. WHEEL "C" HAS A DIAMETER 1/3 THE DIAMETER OF WHEEL "A". WHEEL "A" IS CONNECTED TO WHEEL "C" BY A BELT, AND WHEEL "B" IS FIXED TO WHEEL "C". WHEEL "A" ROTATES IN A CLOCKWISE DIRECTION AT A SPEED OF 10 RPM.

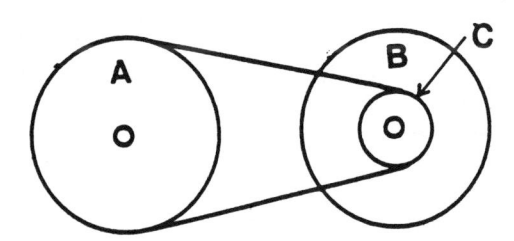

34. If wheel "A" rotates clockwise direction then:

    A. Wheels "C" and "A" turn in opposite directions.
    B. Wheels "B" and "A" turn in the same direction.
    C. Wheels "C" and "B" turn in opposite directions.
    D. Wheels "A" and "B" turn clockwise and "C" turns counterclockwise.

35. If wheel "A" turns at a rate of 10 rpm, the number of rpm made by wheel "C" is most nearly:

    A. 90 rpm.
    B. 60 rpm.
    C. 30 rpm.
    D. 3 rpm.

36. In the drawing below, if wheel #1 is turned in a clockwise direction, wheel #2 will?

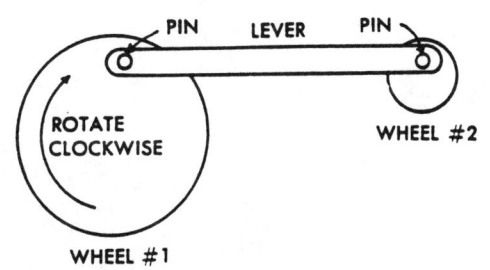

    A. Move back and forth.
    B. Continue to rotate in a clockwise rotation.
    C. Rotate in a counterclockwise rotation.
    D. Wheels #1 and #2 will jam and stop rotating.

37. Of the four following drawings, which one has the weights that will move?

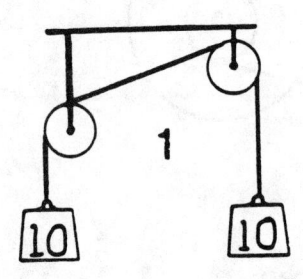

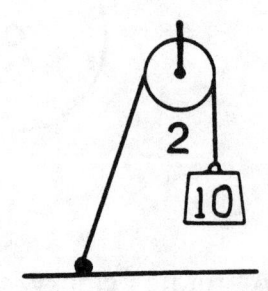

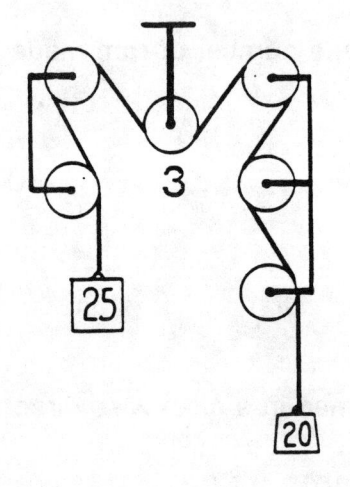

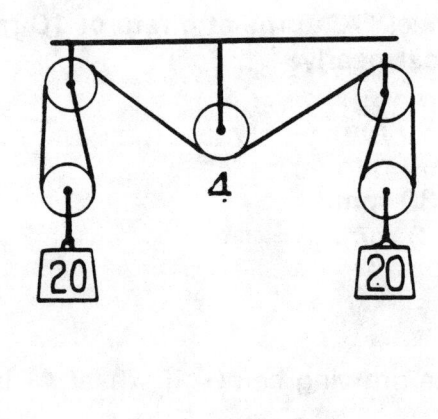

A. Drawing #1.   B. Drawing #2.
C. Drawing #3.   D. Drawing #4.

38. Of the following two drawings, which one has the weights that will move?

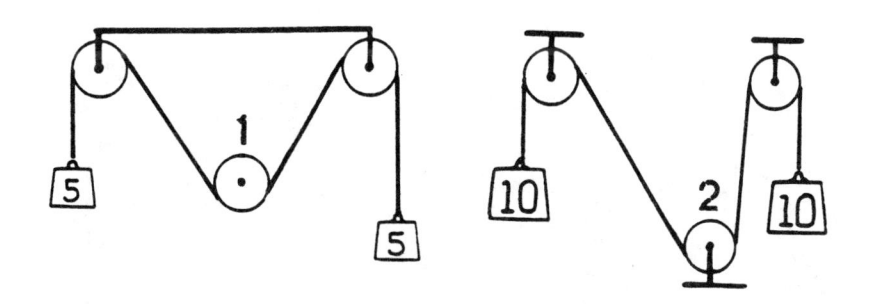

A. Drawing #1.   B. Drawing #2.

39. Of the two following drawings, which one has the weights that will move?

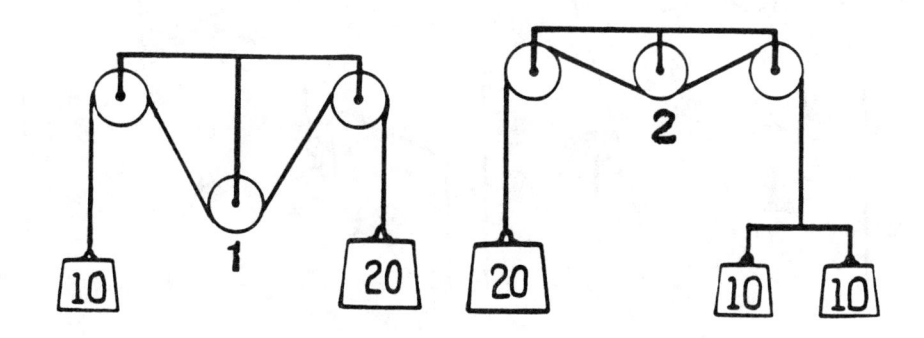

A. Drawing #1.   B. Drawing #2.
C. Both drawing #1 and #2. D. Neither drawing.

40. Of the following four drawings, which one has the weights that will move?

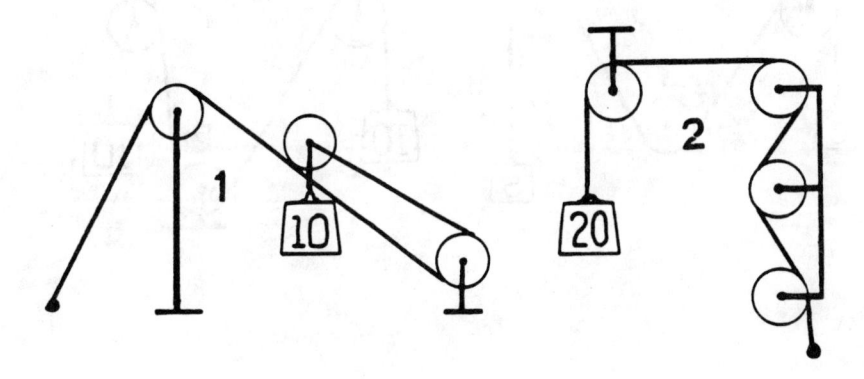

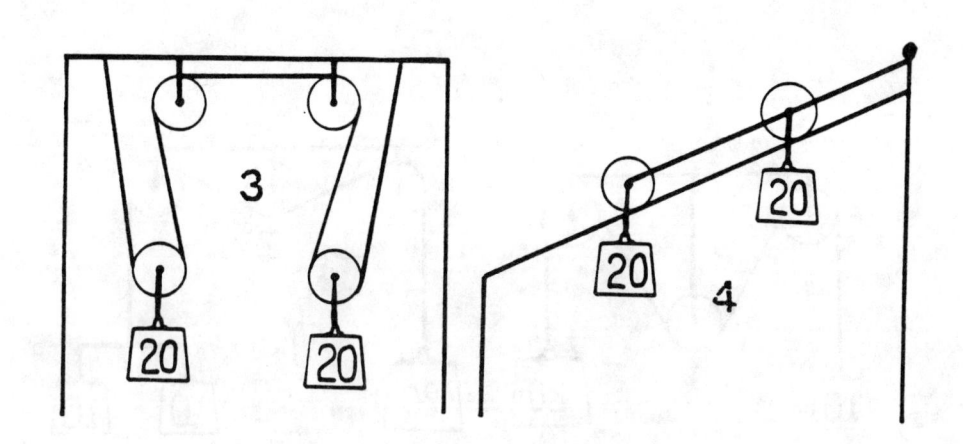

A. Drawing #1.   B. Drawing #2.
C. Drawing #3.   D. Drawing #4.

41. Of the two following drawings, which one has the weights that will move?

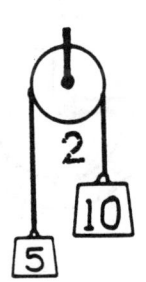

A. Drawing #1.
B. Drawing #2.
C. Both drawing #1 and #2.
D. Neither drawing.

42. Of the two following drawings, which one has the weights that will move?

A. Drawing #1.
B. Drawing #2.
C. Both drawing #1 and #2.
D. Neither drawing.

43. In the drawing below, by crossing the pulley belt:

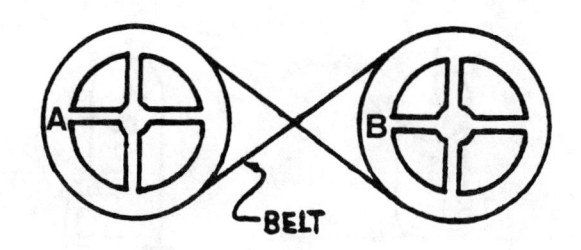

A. The slack will be taken up.
B. The wheels will turn in opposite directions.
C. The belt will last longer.
D. Wheel "A" will turn faster than wheel "B".

44. In the drawing below, the small gear will:

A. Decrease the speed of gears "A" and "B".
B. Increase the speed of gears "A" and "B".
C. Make gears "A" and "B" rotate in opposite directions.
D. Make gears "A" and "B" rotate in the same direction.

45. In the drawing below, pulley "A" rotates in the direction of the arrow, then pulley "B" will rotate:

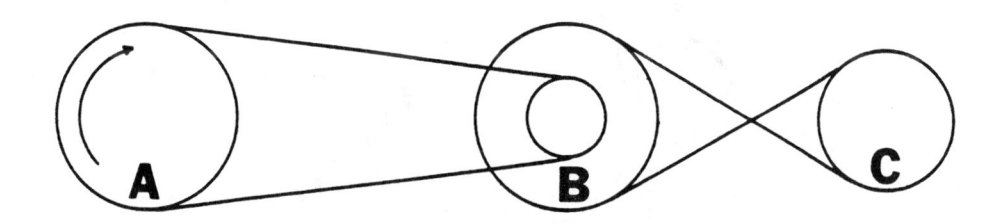

A. Slower than "A", in the opposite direction.
B. Faster than "A", in the opposite direction.
C. Slower than "A", in the same direction.
D. Faster than "A", in the same direction.

46. In the drawing below, if gear "a" rotates in the direction of the arrow, which gear will rotate the fastest?

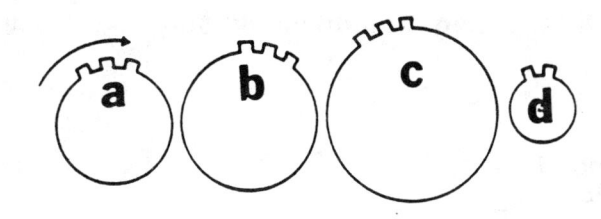

A. Gear "a"
B. Gear "b"
C. Gear "c"
D. Gear "d"

341

QUESTIONS 47, 48, 49, AND 50 ALL RELATE TO THE DRAWING BELOW, WITH GEAR "A" THE DRIVING GEAR ROTATING IN THE DIRECTION OF THE ARROW. GEARS "A" AND "D" HAVE TWICE AS MANY TEETH AS GEAR "B", AND GEAR "C" HAS FOUR TIMES AS MANY TEETH AS GEAR "B".

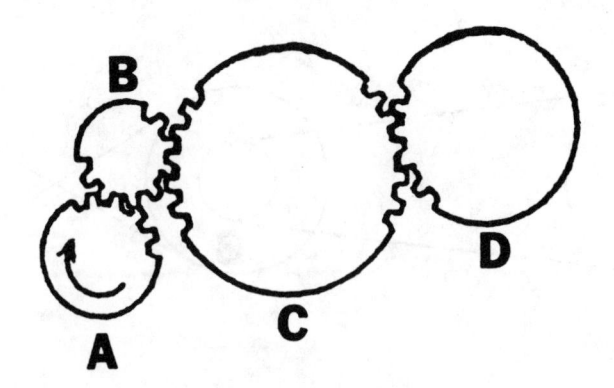

47. Which two gears turn in the same direction?

   A. Gears "A" and "B".
   B. Gears "B" and "C".
   C. Gears "C" and "D".
   D. Gears "B" and "D".

48. Which two gears rotate at the same rate?

   A. Gears "A" and "C".
   B. Gears "A" and "D".
   C. Gears "B" and "C".
   D. Gears "B" and "D.

49. If all of the teeth on gear "C" are knocked off, without influencing the teeth on gears "A", "B" and "D", then rotation would only take place with:

   A. Gear "C".
   B. Gear "D".
   C. Gears "A" and "B".
   D. Gears "A", "B", and "D".

50. If gear "D" rotates at a rate of 100 rpm, then gear "B" rotates at a rate of:

   A. 10 rpm.
   B. 25 rpm.
   C. 50 rpm.
   D. None of the above.

51. The use of an absorbent material placed on a slippery surface, so as to reduce the chances slipping, the affect of an absorbent material is to increase:

   A. Force.
   B. Energy.
   C. Friction.
   D. Gravity.

52. Room temperature is normally higher near the ceiling of a room than near the floor, because:

   A. Hot water pipes are near the ceiling.
   B. The warmer air is lighter than the cooler air.
   C. The air circulates better at floor level.
   D. Most openings are nearer the floor.

53. Two weights of the same size but of unequal weights are dropped at the same time from the same height:

   A. The lightest weight will hit the ground first, because will be greater on the heavier weight.
   B. The heaviest weight will hit the ground first, because it weighs more.
   C. The two weights will hit the ground at the same time, because effect of gravity is the same on both weights.
   D. The two weights will hit the ground at the same time, because they are the same size.

54. The reason it is not appropriate to increase the leverage of a wrench with the use of a pipe over the handle of the wrench, is the wrench:

   A. Will be more difficult to use.
   B. Will be more difficult to put on the nut.
   C. Will slip off the nut.
   D. Could break.

55. The sudden closing of a nozzle on a fire hose that is discharging water under pressure may:

   A. Cause the hose to rupture.
   B. Cause the hose to flail.
   C. Cause the valve to jamb.
   D. Cause the valve to leak.

56. When drilling through wood, you should clamp an additional piece of wood on the underside, to:

   A. Guide the drill bit.
   B. Drill through faster.
   C. Stabilize the drill bit.
   D. Prevent the wood from splintering.

57. To help prevent steel beams from collapsing, during a fire, a layer of concrete may be applied to the steel beams. Why?

   A. Will cause the beams to be stronger during a fire.
   B. Insulates the beams.
   C. Reduces rust and corrosion.
   D. Will cause a chemical reaction during a fire.

58. A fire pump is discharging 250 PSI through 200 feet of fire hose at ground level. When the nozzle is shut off, the pressure at the nozzle =

   A. The same as at the fire pump.
   B. More than the pressure at the fire pump.
   C. Less than the pressure at the fire pump.
   D. More or less than the pressure at the fire pump, depending on the type of fire pump used.

59. Of the following, the most powerful and positive type of clutch is:

   A. Ring.
   B. Flush.
   C. Jaw.
   D. Cone.

60. Of the following valves, the one that should be used either entirely open or fully closed, is:

   A. Globe.
   B. Gate.
   C. Pressure.
   D. reducing.

61. The proper use of the tension pulley on a belt that connects two different size pulleys:

   A. Will not add to the life of the pulley.
   B. Usually positioned closest to the smaller pulley.
   C. Usually positioned closest to the larger pulley.
   D. Usually will be a "V" type pulley.

62. In comparing gasoline engines to diesel engines:

   A. There is less soot produced by diesel engines.
   B. The use of oil additives is not recommended in gasoline engines.
   C. Diesel engines operate at higher temperatures.
   D. Gasoline engines operate at higher temperatures.

63. It is considered poor practice to increase the leverage of a wrench by placing a pipe over the handle of the wrench, because:

   A. The wrench may break.
   B. The wrench may slip from the nut.
   C. It is harder to place the wrench on the nut.
   D. The wrench is more difficult to handle.

64. The type of clutch which gives the most powerful and positive drive is:

   A. Jaw clutch.
   B. Ring clutch.
   C. Disk clutch.
   D. Cone clutch.

65. A tension pulley properly used on a belt connecting two pulleys of different size:

   A. Will commonly be a crowned tension pulley.
   B. Is generally placed nearer the large pulley.
   C. Is generally placed nearer the small pulley.
   D. In no way adds to the belt life.

66. Generally speaking, the practice of racing a car engine to warm it up, is:

    A. Good since repeated stalling of the engine and drain on the battery is avoided.
    B. Good since the engine becomes operational in the shortest period of time.
    C. Bad since proper lubrication is not established soon enough.
    D. Bad since too much fuel is used to get the engine warmed-up.

67. When starting a vehicle equipped with a manual choke, on a cold day, it is best to pull the choke out, because it:

    A. Increases the amount of air in the carburetor.
    B. Reduces the amount of fuel entering the carburetor.
    C. Allows more fuel to enter the carburetor for a richer fuel mixture.
    D. Speeds up the supply of air and fuel to the engine.

68. The best reason for lubricating moving parts of machines is to:

    A. Prevent the formation of rust.
    B. Reduce friction.
    C. Increase inertia.
    D. Reduce the accumulation of dirt on parts.

69. The reason alcohol is added to the radiator of a vehicle in cold weather is because it:

    A. Lowers the freezing point of the mixture.
    B. Lowers the boiling point of the mixture.
    C. Raises the freezing point of the mixture.
    D. Raises the boiling point of the mixture.

70. The water stream from a hoseline that is 100 feet in length has a reach farther than the water stream from a hoseline, from the same pump at the same engine pressure, that is 200 feet long, Why?

    A. The rise of temperature is greater in the longer hose length.
    B. Air resistance to the water stream is proportional to the length of hose.
    C. The time required for water to travel through the longer hose is greater.
    D. The loss due to friction is greater in the longer hose.

71. Firemen usually lean forward when using a charged a charged hose line. What is the best reason for this?

    A. The fireman is more comfortable because of the cooling of the surrounding air at the tip.
    B. A backward force is developed which must be counteracted.
    C. The firemen are better protected.
    D. The firemen can see where the stream strikes better from this position.

72. When water is traveling from a fire stream, and is directed at the roof of a three story building, at what point would the water be traveling at its greatest speed.

    A. As it falls on the roof.
    B. At its maximum height.
    C. At a midway point between the ground and the roof.
    D. As it leaves the hose nozzle.

73. The ammeter of a vehicle will indicate the flow of electrical current:

    A. From the battery to the starter.
    B. Outside the starting circuit.
    C. To the lights.
    D. To and from the battery.

74. If the temperature gauge in a vehicle indicates that the engine is staring to overheat:

    A. Pour cold water in immediately.
    B. Pour hot water in it immediately.
    C. Pour in anti-freeze immediately.
    D. Allow it to cool down.

75. A weight is to be supported from a brace by a chain of 30 links and a hook. Each link of the chain weighs 2 pounds and can support a weight of 1,000 pounds, and the hook weighs 10 pounds and can support a weight of 5,000 pounds, what is the maximum load that can be supported from the hook?

    A. 25,000 pounds.   C. 1,000 pounds.
    B. 5,000 pounds.   D.   930 pounds.

MATCHING TOOLS

MATCH THE PROPER TOOL FROM COLUMN #1 TO BE USED WITH THE PROPER FASTENING DEVICES IN COLUMN #2:

COLUMN #1	COLUMN #2
76.	A.
77.	B.
78.	C.
79.	D.
80.	E.
81.	F.
82.	G.
83.	H.

MATCHING MECHANICAL PARTS

MATCH THE PROPER MECHANICAL PART FROM COLUMN #1 TO AN ASSOCIATED MECHANICAL PART IN COLUMN #2:

COLUMN #1

84.

85.

86.

87.

88.

89.

90.

COLUMN #2

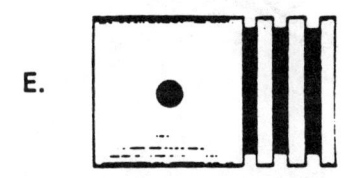

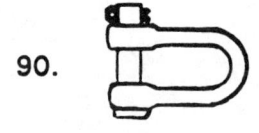

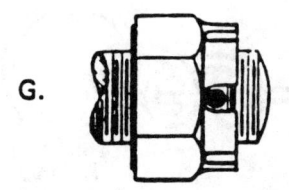

A.

B.

C.

D.

E.

F.

G.

MATCHING KNOTS

COLUMN #1   **COLUMN #2**

91.   A. A bowline.

92.   B. A sheepshank.

93.   C. A clove hitch.

94.   D. Two half hitches.

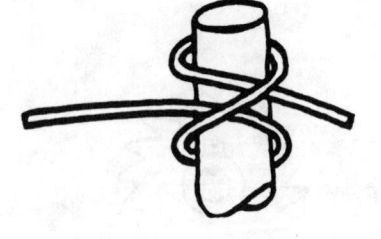

95.   E. A knot which may safely be used for joining two ropes of different sizes.

96.   F. A knot which may safely be used for joining two ropes of the same size, but is not safe for ropes of different sizes.

MECHANICAL OBJECT QUESTIONS

97. The object shown below is a:

    A. Ballpeen hammer.
    B. Framers hammer.
    C. Claw hammer.
    D. tack hammer.

98. The object shown below is a:

    A. T-square.
    B. Depth gauge.
    C. Combination square.
    D. Guide for a saw.

99. The object shown below is a:

    A. Transit knob.
    B. Center punch.
    C. Clock pendulum.
    D. Plumb bob.

100. The object shown below is a:

    A. Drawer pull.
    B. Cleat.
    C. Door stop.
    D. Window lock.

101. The object shown below is a:

    A. Manifold gasket.
    B. Brake band.
    C. Clutch plate.
    D. Air filter.

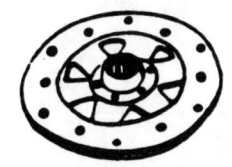

102. The object shown below is a:

    A. paper clip.
    B. wire staple.
    C. Cotter pin.
    D. finishing pin.

103. The object shown below is used to:

    A. Scrape paint.
    B. Etch leather.
    C. Install widow glass.
    D. Drive screws in close areas.

104. The object shown below is used:

    A. For welding.
    B. For soldering.
    C. For lubricating.
    D. For painting.

105. The object shown below is used to:

    A. Measure outside dimensions.
    B. Measure inside dimensions.
    C. Draw circles.
    D. Mark length.

106. The object shown below is used to:

    A. Punch holes.
    B. Clamp objects.
    C. Measure thickness.
    D. Measure depth.

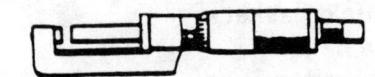

352

107. The object shown below is used to:

   A. Measure depth.
   B. Measure thickness.
   C. Punch holes.
   D. Test metal hardness.

108. The object shown below is used to:

   A. Set spark plugs.
   B. measure saw blades.
   C. Measure drill bits.
   D. Measure wire.

109. The object shown below is used to:

   A. Refill batteries.
   B. Extinguish fires.
   C. Lubricate vehicles.
   D. Kill insects.

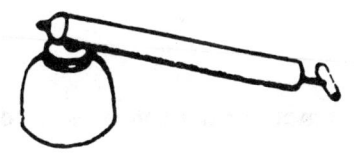

110. The object shown below is used to:

   A. Countersink.
   B. Back screws out.
   C. Remove rivets.
   D. Drill in glass.

111. The object shown below is used to:

   A. Tap pipes.
   B. Turn nuts.
   C. Turn philip screws.
   D. Round corners of nuts.

112. The object shown below is used to:

   A. Start engines.
   B. Replace spark plugs.
   C. Drill holes.
   D. Change auto tires.

113. The object shown below is used to:

   A. Lift automobiles.
   B. Open oil drums.
   C. Straighten bumpers.
   D. Measure heights.

114. The object shown below is used to:

   A. Thread pipes.
   B. Flare metal tubing.
   C. Pull gears off shafts.
   D. Measure diameters.

115. Of the following choices, which tool is the best choice to drill a 3/4 inch hole through a concrete wall:

   A. Jackhammer.
   B. Auger drill.
   C. Star drill.
   D. Diamond point drill.

ANSWERS

1. C	26. A	51. C	76. E	101. C
2. B	27. B	52. B	77. G	102. C
3. C	28. B	53. C	78. B	103. D
4. C	29. D	54. D	79. F	104. A
5. B	30. D	55. A	80. D	105. B
6. C	31. C	56. D	81. H	106. C
7. A	32. A	57. B	82. A	107. A
8. A	33. A	58. A	83. C	108. D
9. C	34. B	59. C	84. G	109. D
10. C	35. C	60. B	85. C	110. A
11. A	36. D	61. B	86. E	111. B
12. A	37. C	62. C	87. F	112. D
13. B	38. A	63. A	88. D	113. A
14. C	39. A	64. A	89. A	114. C
15. B	40. A	65. C	90. B	115. C
16. B	41. B	66. C	91. F	
17. A	42. D	67. C	92. A	
18. B	43. B	68. B	93. A	
19. A	44. D	69. A	94. E	
20. D	45. B	70. D	95. C	
21. A	46. D	71. A	96. D	
22. A	47. D	72. D	97. C	
23. B	48. B	73. D	98. A	
24. B	49. C	74. D	99. D	
25. C	50. C	75. D	100. B	

CHAPTER 12

FIRE SERVICE INFORMATION

INFORMATION SHEET #10

TOPIC: FIRE PUMPS

INTRODUCTION:
The following information is a list of pertanient information concerning fire pumps that all Firefighter candidates should be familair with.

The **COMPOUND GAGES** which are installed on the suction side of fire pumps are calibrated to measure vacuum in inches of mercury and pressure in pounds per square inch. (PSI)

BAROMETRIC PRESSURE will affect the drafting ability of a pump.

The main **FEATURE** of **RELIEF VALVES** are their **SENSITIVITY** to change in pressure and their
ability to relieve this pressure within the pump discharge.

The main **PURPOSE** of the **RELIEF VALVE** which is installed on a pumper engine is to permit water to flow from the discharge to the suction side of the pump.

The main **FEATURE** of a **GOVERNOR** is that it regulates power output of the engine so that it matches pump discharge requirements.

The type of gage most commonly used on the suction side of centrifugal fire pump is the **DOUBLE-SPRING COMPOUND GAGE**.

The capacity of a **POSITIVE-CAPACITY** pump is limited by its displacement and revolutions per minute.

The **POSITIVE DISPLACEMENT** pump is seldom used to produce fire streams in modern times.

For producing fire streams, the **POSITIVE DISPLACEMENT** pump has largely been replaced by the **CENTRIFUGAL** pump.

The most simple **POSITIVE DISPLACEMENT** pump is the **PISTON** pump.

A pump that has a piston with a **BACKWARDS** and **FORWARD** motion is called a **RECIPROCATING PUMP**.

A **POSITIVE DISPLACEMENT PUMP** will have a measurable capacity per stroke revolution.

CENTRIFUGAL means proceeding away from the center, developing outward, or impelling an object outward from a center of rotation.

In a **CENTRIFUGAL** pump the rotating wheel is known as the **IMPELLER**.

Each impeller and housing in a **CENTRIFUGAL** pump is called a "**STAGE**".

As used in the fire department pumping manuals, the term "**NET PUMP PRESSURE**" is best defined as the pressure actually produced by the pump.

1500 GPM is the **HIGHEST STANDARD** size GPM pumper.

TRIPLE COMBINATION PUMPER: hose, water, and pump.

QUAD: hose, water, pump, and ground ladders.

QUINT: hose, water, pump, ground ladders, and aerial ladder.

GAWR: gross axle weight rating.

GCWR: gross combination weight rating.

GVWR: gross vehicle weight rating.

Pumps rated capacity is determined by **TESTING**.

The capacity of a **POSITIVE DISPLACEMENT** pump is limited by its displacement.

If discharge valve is closed on **CENTRIFUGAL** pump the load is decreased on the motor.

VACUUM: space completely void of matter.

HEAT EXCHANGER = "Indirect auxiliary cooler system".

RADIATOR COOLER = "Direct auxiliary cooler system"

CENTRIFUGAL PUMPS are not able to create a vacuum, therefore they need **PRIMING DEVICES**.

CENTRIFUGAL PUMP PRINCIPLE: rapidly revolving disc will tend to throw a liquid from the center toward the outer edge of a disc.

CENTRIFUGAL PUMPS are pumps with one or more impellers that rotate on a shaft, taking water into the eye of the impeller and discharges through the volutes. Centrifugal pumps may be single stage or multiple stage.

In **CENTRIFUGAL PUMPS**: with the quantity remaining constant, the pressure will increase at a rate equal to the square of the speed increase.

In both positive and single stage centrifugal pumps: with the pressure remaining constant, the quantity of discharge is **DIRECTLY PROPORTIONAL** to the speed of the pump.

CENTRIFUGAL PUMP PRINCIPLE: Tendency of revolving body to fly outward from the center of rotation.

CENTRIFUGAL PUMP: water is expelled from one place on the perimeter.

In a **CENTRIFUGAL PUMP**, the pump speed is greater than then engine speed.
POSITIVE DISPLACEMENT PUMP PRINCIPLE: Incompressibility of water (1% to 30,000 lbs pressure)

CENTRIFUGAL PUMPS: Employ a certain principle of force in pumping. The power or force to create pressure is exerted from the center. The revolving motion of the impeller will whirl water introduced at the center toward the outer edge of the impeller. Here it is trapped by the pump casing and is forced to the discharge outlet.

In **CENTRIFUGAL PUMPS**, power is transmitted from the drive shaft, through the pump transmission, and intermediate gear to the impeller shaft.

VOLUTE handles the increasing volumes of water.

VOLUTE is the design of water passageway in a centrifugal pump.

MAIN ACTION of any pump is to add pressure to the water.

VOLUTES primary purpose is to handle an increasing volume of water.

In **POSITIVE DISPLACEMENT PUMP** the pump speed is less than the engine speed. (greater efficiency)

One of the major **DIFFERENCES** between positive displacement pumps and centrifugal pumps in fire fighting apparatus is that centrifugal pumps will not pump air.

With **BOTH** centrifugal and positive displacement pumps, the quantity of water which is discharged is directly proportional to the pump speed when the pressure is constant.

Changing pressure on a single stage centrifugal pump depends on the **MOTOR SPEED**.

The **MAIN ADVANTAGE** that a centrifugal pump has over positive displacement pumps is that the centrifugal pump can exceed its rated capacity.

CENTRIFUGAL PUMPS have the fewest parts to wear out of any fire service type pump.

In a centrifugal pump the quantity of water issuing from the impeller remains **CONSTANT** throughout the entire rotation.

CENTRIFUGAL PUMP: cannot pull a vacuum because it works on the principle of **NON-DISPLACEMENT**.

CENTRIFUGAL PUMPS: are less likely to be damaged by fire pump operators because they only churn water in the within the pump chamber when hose lines are shut down.

In a centrifugal pump, the energy being imparted to the water initially creates **VELOCITY**.

In positive displacement pumps **SLIPPAGE** is dependent on the condition of the pump and the operating pressure.

PARALLEL OPERATION: each impeller discharges into common outlet = increased flow, reduced pressure.
When the transfer valve of a series-parallel pump is placed in the **PARALLEL** position, water from the first stage impeller is routed directly to the pump discharge.

After water has passed through the impeller into the pump volute, it is mainly prevented from returning to the suction side of the pump by the **VELOCITY** of the water.

In origin pump damage from cavitation = **MECHANICAL**. In a series-parallel centrifugal pump the **CLAPPER** check valves are closed by the first stage pressure.

The **TRANSFER VALVE** on fire apparatus is used to allow the pump to operate most effectively.

SERIES-PARALLEL PUMPS are commonly utilized in fire apparatus because they are capable of providing **GREATER VARIATIONS** in capacity and pressure.

Two impellers working in **SERIES** in a two stage pump will give reduced volume at a higher pressure.

A multi-stage centrifugal pump uses two or more impellers to build-up **PRESSURE**.

When pumping from a hydrant supply, the rated capacity of a centrifugal pump is normally **EXCEEDED**.

Compared with positive displacement pumps, centrifugal pumps are generally **LESS EFFICIENT**.

Compared with centrifugal pumps, **ROTARY GEAR PUMPS** generally are more suitable for pumping air.

The object of **PACKING** on a pump is to prevent an air or liquid leak.

The capacity rating given to a pumper is on the basis of delivering full rated capacity from draft at a **10 FOOT LIFT**.

Pump **CAVITATION** occurs mainly in the **IMPELLER EYE**.

A **CLAPPER VALVE** is an automatic valve installed in hydraulic systems, which permits the flow of liquid in **ONE DIRECTION ONLY**.

Only wearing parts of centrifugal pumps are the **BEARINGS**.

VACUUM PUMP: removes air or other gases.

FLUCTUATING PRESSURE is called "**HUNTING**".

CAVITATION can occur whenever a pump is used improperly for existing conditions:
1. Delivering more water than can enter the pump.
2. Excessive lift.
3. Suction hose too small for amount of discharge.
4. Suction blocked at strainer.
5. Suction collapse.
6. Water temperature too high. (above 85 degrees F)
7. Low atmospheric pressure. (high altitudes)
8. A combination of any of one or all of these conditions, at any tank, hydrant, relay, of drafting operation.

PULSATING PRESSURE: indicates air leaks, or restricted suction.

If discharge valves are **CLOSED** on centrifugal pumps, the load is **DECREASED** on the motor

RATED CAPACITIES OF PUMPERS:
1. 2000 GPM.
2. 1750 GPM.
3. 1500 GPM.
4. 1250 GPM.
5. 1000 GPM.
6. 750 GPM.
7. 500 GPM.

When pumping it is safer to stand to the **INSIDE** of the bend in charged hose line.

Pumpers **SHOULD** have service test annually and after major repairs.

Pumps that are rated **LESS THAN 500 GPM** and that are permanently mounted are called **BOOSTER PUMPS**.

The **SMALLEST** size pump recognized for a fire pumper is 500 GPM.

ROTARY VANE hydraulic motors operate both clockwise and counter clockwise.

CERTIFICATION TEST AT DRAFT:
1. 100% volume @ 150 PSI for 2 hours.
2. 100% volume @ 165 PSI for 10 minutes. (spurt test)
3. 70% volume @ 200 PSI for 30 minutes.
4. 50% volume @ 250 PSI for 30 minutes.

DELIVERY TEST (ACCEPTANCE) AT DRAFT:
Same as certification test, plus driving performance, carrying capacity, cooling system, suspension, and braking system. This is the road test. Test should be conducted by joint supervision of representatives from the manufacture and the Fire Department. This is considered the **MOST IMPORTANT TEST**, it is the baseline for later comparisons with the service test.

SERVICE TEST AT DRAFT:
1. 100% volume @ 150 PSI for 20 minutes.
2. 100% volume @ 165 PSI for 5 minutes.
   (not required)
3. 70% volume @ 200 PSI for 10 minutes.
4. 50% volume @ 250 PSI for 10 minutes.

A **MULTI-STAGE** centrifugal pump is equipped with **TWO** or **MORE** impellers.

A two stage of centrifugal pump would most likely have two impellers connected in **SERIES**.

Engine pumper must be able to produce its rated capacity at **80%** or less of its **PEAK** engine speed and must be able to produce its rated pressure at **90%** or less of its **PEAK** engine speed.

The **ACTUAL CAPACITY** of a centrifugal pump is limited by its design: intake diameter, impeller, eye of impeller, diameter of impeller eye, width of impeller, shape and number of vanes in the impeller, and the design of the volute chamber.

TRANSFER VALVE changes pump from volume to pressure, and vice-versa.

SHROUD is the casing of the impeller.
Impellers are made of **BRONZE**. Impeller shafts are made of **STAINLESS STEEL**.

Centrifugal pump **WEAR RINGS** (clearance rings) are positioned nearest the impeller eye.

CENTRIFUGAL PUMP IMPELLER:
1. Large eye   = MORE GPM
2. Wider impeller   = MORE GPM
3. More vanes   = MORE GPM
4. Less vane curvature angle   = MORE GPM

VOLUTE: enables centrifugal pump to handle the increasing quantity of water towards the discharge outlet and at the same time permit the velocity of the water to remain constant or to decrease gradually maintaining the continuity of flow.

WATER UNDER PRESSURE, in the volute of a centrifugal pump is prevented from returning to the suction side of the pump by close fit of the impeller hub to a stationary clearance (wear) ring at the eye of the impeller, and hydraulic pressure due to the velocity created by the centrifugal force.

INFORMATION SHEET #11

TOPIC : FORCIBLE ENTRY/SALVAGE/OVERHAUL/VENTILATION

INTRODUCTION:
The following information is a list of pertanient information concerning forcible entry, salvage, overhaul and ventilation that all Firefighter candidates should be familair with.

Regardless of the class of a door, firefighters should be sure that the door is LOCKED before attempting to gain entry to a building with the use of force.

BEFORE attempting to force any door the firefighters should:
1. Check to see if door is locked.
2. See if hinge pins can be removed.
3. Have hose lines available.

BREACHING: the opening of masonry walls.

Wooden joist of a wood floor are usually a maximum of **16 INCHES APART**.

FORCIBLE ENTRY: entry into a secured building with a minimum of delay, often by the use of special tools. (forcible entry tools).

CLASSIFICATION OF DOORS FOR FIREFIGHTERS:
1. Swinging.
2. Revolving.
3. Overhead.
4. Sliding.

WOOD SWINGING DOORS:
1. Panel.
2. Slab.
3. Ledge.

FORCIBLE ENTRY TOOLS: tools carried on fire apparatus used to gain entry into buildings and obstructions so that firefighting and rescue operations may be carried out.

The **BEST** way for firefighters to learn to recognize various types of windows is to be involved in through building inspection surveys

The distance from the window sill to the floor is usually about **4 FEET**.

As far as **GLASS-PANELED** doors go, it is **CONSIDERABLY** more expensive to replace

TEMPERED plate glass than any other of the same size.

Firefighters should use every other available means of forcible entry before trying to gain entry through a **TEMPERED PLATE GLASS DOOR**.

To prevent glass from sliding down an axe handle while a firefighter is breaking a plate glass window, the firefighter should **STRIKE THE UPPER PART OF THE WINDOW FIRST WHILE STANDING TO ONE SIDE.**

To protect contents of a building and reduce water damage during a fire, use **SALVAGE COVERS** over as much of the building and its contents as possible.

CARRYALL: salvage device that is 6 feet or eight feet square with rope handles at the edges.

FLOOR RUNNER: usually 3 feet by 18 feet (can be up to 30 feet long). Made of canvas or plastic and used on floors to prevent damage to the floor or carpeting from mud, debris, or water.

When VENTILATING a roof, the firefighter should cut a rectangular shaped hole.

For **FORCED AIR VENTILATION** the opening for the replacement air should be equal to or larger than the opening for the venting hole.

SKYLIGHTS that are made from glass are effective vents because the temperature from the fire will break the glass.

WINDWARD: the side of the building where the wind is hitting.

LEEWARD: the side of the building that is opposite the side where the wind is hitting.

INFORMATION SHEET #12

TOPIC : FIRE PREVENTION/BUILDING CONSTRUCTION

INTRODUCTION:
The following information is a list of pertanient information concerning fire prevention and building construction that all Firefighter candidates should be familair with.

U.B.C.: UNIFORM BUILDING CODE is prepared by the International Conference of Building Officials and is published in three year intervals.

U.B.C. : Governs new construction of structures, buildings.

U.F.C. : UNIFORM FIRE CODE is prepared by International Conference od Building Officials and Western Fire Chiefs Association. (technical advisors)

U.F.C. : Governs the maintenance of regulations by Governmental Agencies.

N.B.C. : NATIONAL BUILDING CODE is prepared by A.I.A. which is the American Insurance Association and is published in three year intervals.

N.F.P.A. : NATIONAL FIRE PROTECTION ASSOCIATION is prepared by technically competent committees having balanced representation.

N.F.P.A. : is adopted by public authorities with law making or rule making powers only. (National Fire Codes)

BUILDING CODES are designed to provide rules for public safety in the construction of buildings to the extent which can be applied as law under the broad authority of the police power.

FIRE SAFETY AND CONTROL can best be accomplished by adoption and enforcement of codes and standards.

FIRE LOAD: the expected maximum amount of combustible material in a single fire loss.

STATE GOVERNMENTS do not get involved with forest fire protection.

FEDERAL GOVERNMENT does not get involved with compiling data for insurance and losses.

N.F.P.A. standard calls for automatic smoke vents over theater stages after the "Iroquois" theater fire in **1903**.

FIRE PREVENTION: the fire protection activities with the purpose of preventing fire from starting.

FIRE PREVENTION WEEK: a week devoted to publicizing fire prevention activities. During the week of October 9. (date of great Chicago fire)

Public buildings, schools, hospitals, etc. are all types of **TARGET HAZARDS**.

FIRE PREVENTION BUREAU: a unit of a fire department which does its major work in fire prevention and
investigation rather than combatting fires.

FIVE TYPES OF BUILDING CONSTRUCTION:
    1. Fire resistive.
    2. Heavy timber.
    3. Non-combustible.
    4. Ordinary.
    5. Wood frame.

Buildings with **FIRE-RESISTIVE** construction have the greatest resistance to structural damage by fire.

FOUR TYPES OF WALL CONSTRUCTION:
    1. Reinforced concrete.
    2. Masonry.
    3. Steel frame.
    4. Wood frame.

FIRE RESISTIVE CONSTRUCTION = structural members including walls, partitions, columns, floor, and roof, are made of non-combustible materials of specific ratings.

FIRE DOORS restrict flame, but will allow a great deal of smoke penetration.

STANDARD FIRE DOORS:
    1. Overhead rolling.
    2. Horizontal and vertical sliding.
    3. Single and double swinging.

EXIT DOORS in a theater should swing out in the direction of the street so as the exits are more
readily seen.

AUTOMATIC FIRE DOORS: normally remain open but will close when heat actuates their closing device.

SWINGING FIRE DOOR: used on stair enclosures.

FIRE WALLS: erected to prevent the spread of fire.

FIRE PARTITION is a partition which serves to restrict the spread of fire but does not qualify as a fire wall. (rated for 2 to 4 hours)

BEARING WALL: capable of supporting a vertical load such as a floor, roof, in addition to its weight.

N.F.P.A. table #5 is used for the determination of wall and opening protection.

EXIT ACCESS: the means of egress which leads to an exit.

EXIT: the portion of escape, doors, walls, etc., which provide a protected path to exit discharge.

EXIT DISCHARGE: the portion of travel from the exit to a public way.

MEANS OF EGRESS: continuous, unobstructed way of exit travel from any point in a building or structure to a public way.

The three separate parts of "**MEANS OF EGRESS**":
1. Exit access.
2. Exit.
3. Exit discharge.

THREE CLASSES OF OPENINGS:
1. Class A : Separating buildings or dividing buildings into fire areas.
2. Class B : Vertical openings.
3. Class C : Corridor - room partitions.

CAPACITY OF EXITS is used to establish a consistency of elevation time on the basis of the rate of travel through a door of 60 persons per minute and down a stairway of 45 persons per minute

PANIC HARDWARE on doors, the pressure is not to exceed 15 LBS.

PEOPLE MOVEMENT is the movement of occupants and firefighters.

CLASSIFICATION OF OCCUPANCIES:
1. A = Assembly.
2. B = Business.
3. E = Educational.
4. H = Hazardous.
5. I = Institutions.
6. M = Carports - Fences.
7. R = Residences.

The **TYPE** of occupancy of a building determines the degree and nature of the hazard, along with the fire potential that may exist.

OCCUPANCY HAZARDS are equally as important as construction features in relation to the study of ventilation.

NEVER inspect a building without the permission of the occupant.
LIFE SAFETY CODE requires that in assembly occupancies that the main exit be sized so as to handle at least one half the occupant load. Main exit also serves as entrance.

The **BEST** inspection approach is good public relations, education, and then enforcement.

INSPECTION: the close and critical examination by a competent authority.

Fire prevention inspection should be conducted at **IRREGULAR** intervals so that the inspector may see the inspected establishments in their normal conditions.

NATIONAL FIRE PREVENTION WEEK is declared by **PRESIDENTIAL PROCLAMATION**.

A fire prevention inspection should be the most thorough and complete by the **FIRE DEPARTMENT** than by any other agency.

FIRE HAZARD: any material, condition, or act that will contribute to the start of fire or will increase the severity or extent of the fire.

FIRE CAUSE involves three controllable conditions:
1. Fuel supply.
2. Heat source.
3. The hazardous act.

SPECIAL FIRE HAZARD: fire hazard arising from the processes or operations that are peculiar to the individual occupancy.

COMMON FIRE HAZARD: The condition that is likely to be found in most occupancies and is not generally associated with any specific occupancy, process, or activity.

TARGET HAZARD: A condition, facility, or process which could produce or stimulate a fire that would involve a possible large life loss, a possible large fire loss, a large concentration of materials of high monetary value.

Approximately **99%** of the **SUCCESS** of a fire prevention program depends on **VOLUNTARY** actions of building occupants.

INSPECT DURING FIRE INSPECTIONS:
1. Waste disposal.
2. Trash receptacles.
3. Trash collection points.

RESPONSIBILITY rest with the **FIRE CHIEF** for determining of fire cause and origin.

ARSON is a crime which is hard to secure evidence because the evidence is usually consumed.

MOST IMPORTANT FACTOR in prevention of school fires, is good housekeeping.

SCHOOL INSPECTIONS should be conducted monthly.

Test **HOME** smoke detectors monthly.

SMOKE DETECTORS have the potential to reduce home fires by **40%**. The best spot for **SMOKE DETECTORS** in a home is in the hallway, outside the bedrooms.

E.D.I.T.H. = Exit drills in the home.

Two thirds of all **FIRE DEATHS** occur in the home.

LEADING CAUSE of fire is smoking.

FLAMMABLE LIQUIDS = liquids having flash point below **100 DEGREES F.**

COMBUSTIBLE LIQUIDS = liquids having flash point of **100 DEGREES F OR HIGHER.**

CLASSES OF FLAMMABLE LIQUIDS:
1. Class I: FP = less than 73 degrees F.
2. Class IA: FP = less than 73 degrees F. BP = less than 100 degrees F.
3. Class IB: FP = less than 73 degrees F. BP = greater than 100 Degrees F.
4. Class IC: FP = between 73 and 99 degrees F.

(1-4 FP = flash point, BP = boiling point)

CLASSES OF COMBUSTIBLE LIQUIDS:
1. Class II: FP = 100 to 140 degrees F.
2. Class III: FP = greater than 140 degrees F.
3. Class IIIA: FP = 140 to 199 degrees F.
4. Class IIIB: FP = 200+ degrees F.

INFORMATION SHEET #13

TOPIC : EXTINGUISHING SYSTEMS

INTRODUCTION: As a Firefighter you will be exposed to the various methods of fire extinguishment. The following information pertains to many of the various methods of fire extinguishment and extinguishing systems.

Portable fire extinguisher are classified according to their **INTENDED USE**.

Information for portable fire extinguisher can be found in **N.F.P.A. STANDARD #10**.

FIRE EXTINGUISHER RATINGS AND COLOR CODES:
    Class A fires = Ordinary combustibles: Green triangle.
    Class B fires = Flammable liquids: Red square.
    Class C fires = Electrical: Blue circle.
    Class D fires = Metal: Yellow star.

AGENTS FOR FIRE EXTINGUISHING SYSTEMS:
1. Dry chemical.
2. Carbon dioxide.
3. Foam.
4. Halons.
5. Water spray.

WATER as a **COOLING AGENT** for flammable liquids:
1. Cuts off release of vapor from the surface of a high flash point oil, thus extinguishes the fire.
2. Protects firefighters from flame and radiant heat.
3. Protects flame exposed surfaces.

WATER as a **MECHANICAL TOOL** for flammable liquids:
1. Controls leaks.
2. Directs the flow of the product to prevent it's ignition, or to move the fire to and area where it will do less damage.

WATER as a **DISPATCHING MEDIUM** for flammable liquids:
1. Will float oil above a leak in a tank or during a fire.
2. Will cut off the fuels escape route by pumping it into a leaking pipe before the leak.

WATER is the most important extinguishing agent because of its physical characteristics, its universal availability, and because of it's low cost.

WATER will absorb heat to a much greater extent than any other material that is easily available.

WATER has a lack of opacity, thus it has little ability to prevent the passage of radiant heat.

Adding **WETTING AGENTS** to water increases the waters heat absorption ability.

WETTING AGENTS reduce the surface tension of water.

WETTING AGENTS increase the waters penetration ability.

A gallon of water produces a maximum of **200 CUBIC FEET OF STEAM**.

The formation of steam by water that has been applied to the burning source will temporarily cause an inert gaseous water zone (steam) in and around the burning zone.

WET-WATER solutions will foam easily, and the temporary foam will control and extinguish class B fires better than ordinary water.

BASIC TYPES OF FOAM FOR FIREFIGHTING:
1. Chemical foam.
2. Mechanical foam.

SUBSURFACE FOAMS injected or intermixed with the liquids that are involved are called flouroprotein.

WETTING AGENTS reduce the surface tension of water which increases the penetrability.

SURFACE TENSION is the resistance to penetration possessed by the surface of a liquid.

DIRECT foam against the far side of tank shell when extinguishing an oil tank fire.

The chief purpose of a **STABILIZER** in foam extinguisher is to prevent the rapid breakdown of the carbon dioxide (CO_2) bubbles.

AFFF FOAM RATIO:
1. Hydrocarbons (petroleum products) = 3% foam to 97% water.
2. Polarsolvents (water soluble)   = 6% foam to 94% water.

FOAM is the most effective extinguishing agent for oil fires, because it is lighter than oil and remains on top.

WATER FOG is not capable of extinguishing flammable liquid fires with a flash point below 100 degrees F

SPRAY FIRE STREAMS IN CONTRAST TO SOLID STREAMS:

    ADVANTAGES:

1. Absorbs more heat, more rapidly.
2. Covers a greater area with water.
3. Uses less water.

    DISADVANTAGES:

1. Requires higher discharge pressure.
2. Has a shorter reach.
3. Has less penetration.
4. Has less cooling effect in subsurface areas such as charred wood.

VISCOUS WATER (thickened water:
   Advantages
   1. Sticks to burning fuel.
   2. Spreads itself in continuous coating.
   3. Thicker than plain water.
   4. Absorbs more heat.
   5. Projects longer and higher straight streams
   6. Seals fuel from oxygen after drying.
   7. Resist wind drift. (as from aircraft in forest fires).

LIGHT WATER (fluorinated surfactant), is useful in obtaining quick knockdown of flammable liquid fires, and in providing a vapor sealing effect for reducing subsequent flashover of fuel vapors exposed to lingering.

CONVENTIONAL FOAM is formed by the reaction of alkaline salt solution in acid salt solution in the presence of a foam stabilizing agent, and mechanical or air foam formed by turbulent mixing of air with water containing foam forming agents.

HIGH EXPANSION FOAM is for **CLASS "A" AND "B"** fires and suited as a flooding agent in confined spaces.

LOW EXPANSION FOAMS can be used to good effect when the temperature of the bulk of a contained liquid does not exceed **250 DEGREES F.**

ORDINARY FOAM IS BROKEN DOWN BY:
   1. Common alcohols.
   2. Aldehydes.
   3. Ethers.

ALCOHOL FOAM is recommended for all water soluble flammable liquids except for those that are only very slightly soluble.

The **MOST IMPORTANT** use of **FOAM** is in fighting fires in petroleum hydrocarbons (gasoline) with high vapors and low flash points.

FOAM should never be used on energized electrical equipment.

PROPORTIONER is the device that is used to inject the correct amount of foam concentrate into the water stream so as to make the foam solution the correct proportion.

FIVE GALLONS of foam powder = **400 GALLONS** of foam.

2A WATER fire extinguisher will cover a **100 FOOT** area.

The principle extinguishing agent of **SODA** and **ACID EXTINGUISHER** is the **WATER** content.

The most widely used water soluble freezing point depressant in fire equipment is **CALCIUM CHLORIDE.**

SODA ACID EXTINGUISHER have Bicarbonate Soda and Sulfuric Acid as ingredients.

REGULAR DRY CHEMICAL EXTINGUISHER are Bicarbonate base powder for Class B and Class C fires. (Sodium Bicarbonate and Potassium Bicarbonate)

MULTIPURPOSE DRY CHEMICAL EXTINGUISHER are Ammonium Phosphate base powders for Class A, B, and C fires. (Monoammonium Phosphate, also Barium Sulfate)

EXTINGUISHMENT BY SEPARATION cannot be accomplished with burning materials that contain their own oxygen supply, such as Cellose Nitrate.

EXTINGUISHMENT BY SEPARATING the oxidizing agent from the fuel is accomplished by blanketing or smothering a fire.
DRY CHEMICAL EXTINGUISHER are a mixture of Sodium Bicarbonate, Potassium Bicarbonate, or Ammonium Phosphate.

DRY CHEMICAL EXTINGUISHER EXTINGUISH BY:
1. Smothering.
2. Cooling.
3. Radiation shielding.
4. Chain breaking; a reaction in the flame may be the principle cause of extinguishment.

CHEMICAL COMPOSITION of most ordinary combustible solids consist of primarily: Carbon, Hydrogen, and Oxygen.

Use **DRY CHEMICAL EXTINGUISHER** on LP gas fires.

DRY CHEMICAL EXTINGUISHER should not be used where relays or other delicate electrical contacts are located such as telephone exchanges.

DRY CHEMICAL EXTINGUISHER will not extinguish fires in materials that supply their own Oxygen.

MULTIPURPOSE DRY CHEMICAL EXTINGUISHER, extinguish by the Ammonium Phosphate decomposing and leaving a sticky residue on the burning material, this seals the material off from Oxygen.

5 LBS of **DRY CHEMICAL** is as effective as **10 LBS** of **CARBON DIOXIDE**.
MULTIPURPOSE DRY CHEMICAL EXTINGUISHER may be used on Class A fires.

FOAM EXTINGUISHER must be protected from freezing.

CARBON DIOXIDE FIRE EXTINGUISHER extinguish by cooling with the rapid expansion of liquid to gas producing a refrigerating effect.

15 LBS CARBON DIOXIDE EXTINGUISHER range = 8'-10'.

CARBON DIOXIDE EXTINGUISHER are not effective on fires involving reactive metals.

CARBON DIOXIDE EXTINGUISHER are under a pressure of between **800 PSI** and **900 PSI**.

CARBON DIOXIDE EXTINGUISHER with CO_2 under pressure, the purpose of the horn dispenser is to avoid entraining air as the contents pass through the small orifice at high velocity.

CARBON DIOXIDE EXTINGUISHER suppress chain reaction of combustion.

1 LB of **CARBON DIOXIDE** liquid will produce **8 CUBIC FEET OF GAS** at atmospheric pressure.

CARBON DIOXIDE EXTINGUISHER are not effective on fires involving chemicals containing their own Oxygen supply, such as Cellulose Nitrate.

CARBON DIOXIDE EXTINGUISHER are recommended for stove fires.

CARBON DIOXIDE EXTINGUISHER are recommended for fires involving energized electricity.

Use care with **CARBON DIOXIDE EXTINGUISHER**, mainly because of the possibility of **RE-FLASH**.

CARBON TETRACHLORIDE vaporizes to form a heavy non-flammable gas.

PORTABLE FIRE EXTINGUISHER WHICH ARE CONSIDERED OBSOLETE:
1. Soda Acid.
2. Carbon Tetrachloride.
3. Loaded Stream.
4. Cartridge-operated Water.
5. Inverting Foam.

COMBUSTIBLE METAL FIRES (Class D) are extinguished by a number of approved extinguishing agents; Powders, and Dry Powders. Each for specific metals.

HALOGENATE EXTINGUISHING agents work by:
1. Vaporizing liquid fire extinguishing agents.
2. Chain Breaking.
3. By providing non-flammability and extinguishing characteristics.

HALON EXTINGUISHING AGENTS:
1. Fluorine.
2. Chlorine.
3. Bromine.
4. Iodine.

SPRINKLER RATINGS:
Ordinary	- No color	= 135	- 170 degrees F.
Intermediate	- White	= 175	- 225 degrees F.
High	- Blue	= 250	- 300 degrees F.
Extra high	- Red	= 325	- 375 degrees F.
Very X high	- Green	= 400	- 475 degrees F.
Ultra high	- Orange	= 500	- 575 degrees F.

SPRINKLER COVERAGE according to occupancy hazard, with 1/2" orifice:
Light = 130 square feet to 168 square feet.
Ordinary = 130 square feet.
Extra high = 90 square feet.

WATER SUPPLIES FOR SPRINKLER SYSTEMS:
1. Public water works systems.
2. Public and private supplies.
3. Gravity tanks. (minimum of 5000 gallons)
4. Pressure tanks. (minimum of 4500 gallons)
5. Fire pumpers.
6. Fire department connections.

AUTOMATIC SPRINKLERS should be installed in warehouses of Type I or II, but not of III.

The **MAIN** reason that a sprinkler system should be installed inside of buildings is to extinguish fires in their early stages.

AUTOMATIC SPRINKLERS under normal situations are highly efficient and dependable. If an explosion takes place, sometimes this will cause the sprinkler head and piping to be so badly damaged that they will be ineffective.

AUTOMATIC SPRINKLERS are the most effective safeguard against loss of life by fire. Psychological as well as physical, by minimizing panic.

TOTAL SPRINKLER SYSTEMS (piping and devises) are hydrostatically tested at not less than **200 PSI FOR 2 HOURS** or 50 PSI in excess of the maximum static pressure when it is above 150 PSI.

PRE-ACTION SPRINKLER SYSTEMS are activated at the water supply valve, not at the sprinkler head, like in the standard type dry sprinkler system.

DRY PIPE SPRINKLER SYSTEMS, the piping contains air under pressure, 15 PSI to 20 PSI and no water in the piping.

WET PIPE SPRINKLER SYSTEMS are fully charged with water.

DELUGE SPRINKLER SYSTEMS wet down the entire area by admitting water to sprinklers that are open at all times.

OUTSIDE SPRINKLER SYSTEMS, the water curtain is for exposed protection.

AUTOMATIC SPRINKLER SYSTEMS extinguished or held in check 96% of fires in which they were involved.

SPRINKLER PATTERN will equal **16 FOOT DIAMETER** circle at a point 4 feet below the sprinkler head at 15 GPM. Sprinkler head areas will **OVERLAP**.

Sprinkler systems are to be inspected **4 TIMES PER YEAR**.

Maximum number of sprinkler heads that any be supplied through a single Deluge valve is **5 SPRINKLERS** per 1 1/2 inch valve.

PRE-ACTION SPRINKLER SYSTEMS ARE ACTIVATED BY:
    1. Smoke detectors. and 2. Heat sensors.

"PINTLE" on sprinkler head indicates that the orifice is smaller than the standard 1/2 inch size.

ACCELERATORS and **EXHAUSTERS** speed up the expelling of air from Dry pipe systems.

TYPES OF STANDPIPE SYSTEMS:

1. Wet, supply valve open with water pressure at all times.
2. Dry, no permanent water supply in sprinklers.
3. Automatic supply system, opening hose valve.
4. Manual to remote system, is at hose station.

CLASS I STANDPIPE SYSTEM: For use by Fire Departments and those trained in handling heavy fire streams. (2 1/2 inch hose inside building or for exposure.)

CLASS II STANDPIPE SYSTEMS: For use primarily by the building occupants until the Fire Department arrives. (1 1/2 inch hose for incipient fires.)

CLASS III STANDPIPE SYSTEMS: For use by either Fire departments and those trained in handling heavy hose streams or by the building occupants on small hose streams. (for large or small fires.)

OS&Y VALVE:
1. Outside Screw and Yoke Valve.
2. Outside Stem and Yoke Valve.

INDEX

INDEX

A

ABILITIES 2, 10, 11, 13, 40, 64, 155, 161, 302, 10
ACTING POSITION 4
ADDITION   41, 63, 76, 176, 177, 187, 205, 253, 271, 368
ADIABATIC 287
ADMINISTRATIVE SUPPORT 4, 5
AERIAL LADDER 13, 14, 22, 23, 26, 359, 23
AFFF 372
AFFIRMATIVE ACTION 41
AFTER THE EXAM 78
AFTER THE ORAL INTERVIEW 163
AGE 9
AIR CLEANER 310
ALCOHOL FOAM 373
ALPHABETICAL PROGRESSIONS 68
ALTERNATOR 308
AMBULANCE DRIVER 60
AMMETER 308
AMPERES 286, 303
ANGLE 285, 286, 308, 363, 285
ANNOUNCEMENT 39, 41, 45, 47, 49
ANSWERING QUESTIONS 160
APPARATUS . 4-6, 10, 11, 13-19, 21, 22, 26, 27, 29, 36-38, 58, 60, 61, 81, 89, 97, 100, 103, 104, 110, 114, 115, 170, 171, 197, 225, 248, 250, 251, 254, 267, 268, 273, 304-313, 360, 361, 364, 310
APPARATUS DRIVER 4, 19, 81, 100, 103, 104, 110,114, 312
APPLICANTS 11
APPLICATION 3, 11, 12, 24, 27-29, 33-35, 39, 42, 48-50, 52, 55, 65, 78, 155, 157-159, 161, 163, 164, 245, 282
APTITUDE .. 11, 62, 68, 79, 90, 99, 176, 299, 315, 317-319, 11
AREAS 2, 3, 11, 13, 19-22, 26, 27, 41, 42, 55, 78, 156, 176, 207-209, 230, 248, 262-264, 278, 282, 288, 314, 318, 352, 368, 372, 376
ARITHMETIC .. 68, 76, 175-177, 76, 176
ARSON 5, 15, 36, 58, 119, 294, 369
ASSESSMENT 36, 157, 174
ASSISTANT CHIEF 4
ATMOSPHERIC PRESSURE .. 93, 279, 280, 284, 286, 362, 374
ATTITUDE ... 12, 132, 155, 161, 163, 12
AUTOMATIC SPRINKLER SYSTEMS 305, 376
AUTOMOTIVE EQUIPMENT 27
AUXILIARY COOLER 359

B

B.T.U. 93, 279
   BRITISH THERMAL UNIT 279
BACK PRESSURE 252, 253, 284
BACKGROUND ASSESSMENT 174
BACKGROUND INVESTIGATION 11
BAROMETRIC PRESSURE 358
BASIC DRILLS 27

B
(continued)

BASIC HYDRAULICS 207, 208, 220
BASIC SCIENCE 277, 278
BATTERIES 15, 307, 313, 353, 307
   CHARGE 308
   CHARGING RATE 307
   ELECTROLYTE 307
   OVERCHARGING 308
   SPECIFIC GRAVITY 307
   SULFURIC ACID 307
BELTS 103, 114, 250, 302, 306, 314
BENEFITS 3, 7, 42, 116, 7
BEST FOOT FORWARD 161
BIMETALLIC 286
BOILING POINT 73, 215, 280, 281, 309, 313, 346, 370
BOWLINE KNOT 306
BRAKES 307
   EXCESSIVE HEAT 307
   FADING 307
   FLUID 307
   LINING 307
BREATHING APPARATUS . 14, 15, 21, 26, 103, 114, 254, 306
BRITISH THERMAL UNIT 98, 279
   B.T.U. 279
BUILDING CONSTRUCTION   11, 13, 36, 57, 61, 63, 366, 367, 366
BURNS 242, 245, 255, 282, 290

C

CADET FIREFIGHTER 3
CALL FIREFIGHTER 3
CALORIE 93, 287
CAM FOLLOWER 310
CAM SHAFT 310
CAMBER 75, 309, 312
CAPACITIES 207-209, 271, 362
CAPACITOR 308
CARBON BLACK 310
CARBON DIOXIDE ... 245, 247, 260, 261, 291, 371, 372, 374, 375
CARBON DIOXIDE EXTINGUISHER 374, 375
CARBON MONOXIDE ... 94, 98, 274, 289, 291, 295, 296
CARBURETOR 309, 310
CAREER CLASSIFICATIONS 3
CAREER OPPORTUNITIES 33-35
CAREERS/OPTIONS 2
CARRYALL 365
CASTER 309
CAVITATION 361, 362
CENTER OF GRAVITY 314
CENTIGRADE 212, 215, 221, 240
CENTRIFUGAL 358
CENTRIFUGAL FORCE 270, 285, 363
CENTRIFUGAL PUMP 269, 270, 358-361, 363
CERTIFICATES 57
CERTIFICATION TEST 251, 362, 363

C
(continued)

CHAIN SAW 26, 305
CHARACTER 11, 12
CHEMICAL REACTIONS 262, 282
CHEMISTRY 11, 68, 69, 93, 225, 277-279, 288, 289, 298, 299, 318
CHEMISTRY AND PHYSICS QUESTIONS 93
CHIEF ENGINEER 4
CHIEF FIRE OFFICER 4
CHOKE 310
CHOKE VALVE 310
CHRONOLOGICAL RESUME 56
CIRCLES 209, 352
CIRCUMFERENCE 183, 210, 213, 220, 285
CLASS C FIRE 245
CLOSING DATE 49
CLOSING STATEMENT 174
CLOVE HITCH 350
CLUTCH 309
COIL: 309
COMBUSTIBLE LIQUIDS 245, 246, 280, 281, 370
COMBUSTIBLE METALS 245-247
COMMITMENT LIST 35
COMMUNICATE 39
COMMUNITY COLLEGE 42
COMPOUND GAGES 358
COMPRESSION 309, 311
COMPRESSION GAUGE 309
CONCENTRATION/MEMORY 236, 239, 275
CONDENSER 308
CONDUCTOR 286
CONVECTION 94, 98
CONVENTIONAL FOAM 373
CONVERSIONS 208, 212
COOLER RADIATOR 359
COOPERATION 27
CORROSIVE CHEMICALS 73
COURAGE 12
COVER LETTER 34, 39, 62-65
CURRENT REGULATOR 308
CURRICULUM 42

D

DECIMAL NUMBERS 178, 179
DEGREE IN FIRE SCIENCE 58
DEPARTMENT FAMILIRIZATION 22
DEPENDABILITY 12
DEPUTY CHIEF 4
DIAGNOSTIC TEST 80, 228, 233, 289, 298, 318, 319
DIESEL ENGINES:TURBOCHARGED ... 311
DISCOVER 40
DISPATCH OFFICER 5
DISPLACEMENT 358
DISPLACEMENT PUMP 358-360
DISTRIBUTOR 309
DIVISION PROBLEMS 193
DIVISION CHIEF 4
DOCUMENTATION 12
DOORS/EXITS 367
DRILLS 27, 29
DRY CHEMICAL EXTINGUISHER 263, 265, 373, 374
DWELL ANGLE 308

E

E.D.I.T.H. 370
E.M.T. PROCEDURES 23
EDUCATION 5, 6, 9, 11, 28, 29, 34, 36, 42, 48, 55-61, 63, 65, 157, 168, 369, 9, 36, 58, 59
EDUCATION/TRAINING 36
EFFICIENT READING 225
EGRESS 368
ELECTRICAL 72, 74, 245, 286, 292, 293,295, 303, 304, 307, 308, 313, 347, 371, 373, 374
ELECTRICAL SHOCK 304
ELECTRICITY 303
ELECTRON 287, 291, 297, 303
ELEMENTS 287
EMERGENCY RESPONSES 6, 7
EMERGENCY WORK 28, 29, 28, 29
EMT-1 .... 10, 57, 59, 57
ENGINES
 4 STROKE 311
 DIESEL 311
 GASOLINE 311
ENTRANCE LEVEL FIREFIGHTER 20
EQUIPMENT 305
ESPRIT DE CORP 6, 56, 62, 6
ESSAY TEST 71, 73
EVALUATING 21
EVALUATIONS 21, 22, 30
EXAM CHECK LIST 39, 157
EXAM HINTS 69
EXAM PREPARATION 39
EXAMS/VOCABULARY 130
EXCEEDED 361
EXERCISE 39
EXIT 368
EXPERIENCE 10, 11, 13, 16, 19, 27, 28, 36, 40, 55-57, 59-64, 144, 155, 161, 162, 168, 229, 262, 10, 35, 57
EXPLORER 3, 8
EXPOSURES 259, 262, 267, 268
EXTINGUISHER 262, 373
 CARBON DIOXIDE 374, 375
 DRY CHEMICAL 373, 374
 DRY POWDER 374
 FOAM 260, 374
 HALON 265
 MULTIPURPOSE 374
 OBSOLETE 375
 SODA ACID 373
 WATER 371
EXTINGUISHER RATINGS 371
EXTINGUISHER/LIQUIFIED 265
EXTINGUISHERS .. 14-16, 21, 24, 73, 245
EXTINGUISHING AGENTS   245, 247, 375
EXTINGUISHING SYSTEMS 19, 371

F

FACTOR RATING 29
FAHRENHEIT 72, 93, 212, 215, 221, 240, 291, 297
FAMILIARIZATION 37
FILL IN THE BLANKS 71, 72
FIRE
 CAUSE 369
 CLASSIFICATIONS 245
 EXTINGUISHER 371

381

F
(continued)

FIRE (continued)
- HAZARD 369
- HOSE 248, 249
- HYDRAULICS 283
- LOAD 366
- POINT 280
- PUMPS 358
- STRATEGY 267
- TRIANGLE 274

FIRE ACADEMY . 3, 11, 101, 111, 248, 11
FIRE BEHAVIOR 19, 21, 25, 61, 267
FIRE CHIEF 4
FIRE DEPARTMENT FAMILIARIZATION . 37
FIRE DEPARTMENT POLICY 21
FIRE DEPARTMENT PUMPERS .. 271, 272
FIRE ENGINEER 4, 14, 18, 19, 60, 127, 173, 204, 305
FIRE EXTINGUISHER 73, 249, 260-263, 371, 373-375
FIRE EXTINGUISHER RATINGS 371
FIRE HOSE ... 68, 69, 77, 115, 200, 210, 248-253, 295, 343, 344
FIRE HYDRANT 15, 82, 271
FIRE HYDRAULICS .. 19, 36, 58, 61, 283
FIRE LIEUTENANT 4
FIRE POINT 93, 280
FIRE PREVENTION 3, 5, 6, 13, 14, 16-19, 21, 24, 25, 28, 29, 35, 36, 58, 61, 69, 77, 83, 244, 274, 366, 367, 369, 24, 28, 29, 366
FIRE PREVENTION BUREAU 5
FIRE SCIENCE 41, 42
FIRE SERVICE ACADEMY 36
FIRE SERVICE CAREERS 2
FIRE SERVICE COURSES 36
FIRE SERVICE DIRECTORIES 41, 44
FIRE SERVICE INFORMATION ... 226, 238
FIRE SERVICE POSITIONS 4
FIRE STRATEGY 267
FIRE STREAMS 21, 271, 272, 310, 358, 372, 377
FIRE SUPPORT SERVICES 5
FIRE SUPPRESSION 4, 13, 64
FIRE TACTICS 58, 267
FIRE TECHNOLOGY COURSES . 41, 42, 58
FIRE TECHNOLOGY DEPARTMENTS .. 42
FIRE TETRAHEDRON 93, 274
FIRE TRIANGLE 93, 98, 274
FIRE: MODES/STAGES 242
FIRE-RESISTIVE 367
FIREFIGHTER ENTRY LEVEL 78
FIREFIGHTER EXAMS 65, 73, 120, 132, 176, 224, 278, 288
FIREFIGHTER ORAL INTERVIEW 154
FIREIGHTER DUTIES 244
FIREFIGHTERS ENTRANCE LEVEL EXAM 68
FIREFIGHTERS WRITTEN EXAM 70
FIREFIGHTING 3, 4, 11, 13-18, 35, 58, 60, 61, 73, 76, 77, 101, 111, 267, 273, 284, 293, 302, 304, 364, 372
FIRES/METAL 375
FIRST AID ... 13, 15-17, 36, 58, 68, 169
FLAMMABLE DENSITY 280
FLAMMABLE LIQUIDS 74, 245, 246, 262, 263, 280, 281, 370, 371, 373

F
(continued)

FLASH POINT . 93, 98, 280, 281, 370-372
FLASHOVER 373
FLUID PRESSURE 213
FOAM 148, 151, 245, 247, 249, 260, 261, 371-375, 372
- AFFF 372
- ALCHOL 373
- COMVENTIONAL 373
- HIGH/LOW EXPANSION 373
- ORDINARY 373
- PROPORTIONER 373

FOAM EXTINGUISHER 372, 374
FORCIBLE ENTRY 21, 24, 255, 288, 364, 365, 24, 364
FORMULAS 208, 211, 230
FRACTIONS 176-178
FRICTION LOSS 251, 283, 284, 303
FUEL 310, 311
- GASOLINE TO AIR MIXTURE 311
- OCTANE 310
FULCRUM 285
FUNCTIONAL RESUME 56
FUSES 308

G

GAGES 213, 358
- COMPOUND 358
GASES 14, 16, 111, 114, 240, 241, 245-247, 254, 259, 268, 279, 280, 282, 289-291, 296, 307, 310, 362
GASOLINE ENGINES 311
GAUGES 213, 271, 272
GAWR 359
GCWR 359
GEAR 309
GENERAL KNOWLEDGE 68, 78, 224, 278, 288, 318
GENERATOR 308
GOVERNOR 358
GRAMMAR 48, 68, 78, 140, 141, 145, 146, 150, 162, 163, 140
GRAMMAR TEST 141, 145, 146, 150
GRAVITY 285
GVWR 359

H

H2O 73
HALOGENATE 375
HALON 245, 247, 265, 266, 375
HALYARD 306
HAZARD:
- FIRE 369
- LIGHT 262
- TARGET 369
HAZARDOUS MATERIALS . 19, 38, 58, 61
HAZARDS
- OCCUPANCY 369
- EXTRA 263
- ORDINARY 262
HEAD PRESSURE 252, 253, 284
HEAT
- EXCHANGER 359
- LATENT 279
- SPECIFIC 279
HONESTY 12

382

H
(continued)
HOSE
    CLAMPS 305
HURST TOOL 23, 255, 256, 255
HYDRAULICS v, 19, 36, 58, 61, 207, 208, 215, 219, 220, 225, 283, 293, 283
HYDROMETER 285, 308

I
IGNITION 308, 311
    SPONTANEOUS   258
    TEMPERATURE 280
IGNITION COIL 308
IGNITION SWITCH 308, 313, 308
IGNITION TEMPERATURE 261, 280
INFORMATION
    GENERAL 303
    MECHANICAL 307
    PHYSICS 285
    RELATING TO FIRE SERVICE 279
INFORMATION SERVICES . . 41-44, 41, 42
INFORMATION SHEET #1 9
INFORMATION SHEET #2 13
INFORMATION SHEET #3 18
INFORMATION SHEET #4 20
INFORMATION SHEET #5 35
INFORMATION SHEET #6 76
INFORMATION SHEET #7 99
INFORMATION SHEET #8 120
INFORMATION SHEET #9 132
INFORMATION SHEET #10 358
INFORMATION SHEET #11 364
INFORMATION SHEET #12 366
INFORMATION SHEET #13 371
INITIATIVE 12, 28, 127, 12
INSPECTION . . . 5, 16, 24, 25, 28, 29, 99, 101, 105, 109-112, 116, 364, 369
INSPECTIONS: SCHOOL 370
INSULATOR 286
INTEGRITY 12, 62, 120, 155, 12

J
JAWS OF LIFE 255
JOB ANNOUNCEMENTS 45
JOB APPLICATIONS 48
JOB KNOWLEDGE 28
JOB OBJECTIVE 59
JOB SECURITY 7, 28, 7
JOB SKILLS 57
JUDGMENTS 77

K
KNOTS 21, 24, 350
    MATCHING 350
KNOW THE COMPETITION 39

L
LADDER: RAISING 273
LADDER COMPANY 111
LADDER TRUCK 13, 14, 26, 204
LADDERS 13-17, 21, 23, 24, 68, 69, 81, 95, 249, 250, 271, 273, 359, 23
LATENT HEAT 73, 94, 98, 279

L
(continued)
LEARN 35
LEEWARD 365
LEVERAGE 285, 343
LIBRARY 35, 42, 203, 237, 315
LICENSES 10, 58
LIFE SAFETY CODE 369
LIFEGUARD 59
LIGHT 287
LIGHT WATER 373
LIQUID MEASURE 216
LIQUIDS
    COMBUSTIBLE 280, 370
    FLAMMABLE 280, 370
LOCATING EXAMS 41
LUBRICANTS 312

M
MAGNETIZATION 286
MAGNETO 308
MAINTENANCE 27
MANIPULATIVE 20, 25, 26
MANIPULATIVE FINAL EXAM 26
MANPOWER 38, 89, 268
MARITAL STATUS 55
MATCHING
    KNOTS 350
    MECHANICAL PARTS 349
    TOOLS 348
MATCHING QUESTIONS 74
MATH
    ADDITION 187
    CONCEPTS 176
    DIVISION 193
    MULTIPLICATION 191
    PERCENTAGE 195
    SUBTRACTION 189
    WORD PROBLEMS 197
MATH SKILLS 184
MATHEMATICS 68, 87, 185, 187
MEALS 6, 7, 16, 200
MEASUREMENT OF AREAS 209
MECHANICAL
    INFORMATION 302, 307
    OBJECT QUESTIONS 351
    PRINCIPLES 313
MECHANICAL APTITUDE 62, 90, 315, 318, 319
MECHANICAL APTITUDE QUESTIONS . . 90
MECHANICAL COMPREHENSION . 68, 302
MECHANICAL PARTS 349
MEDICAL 9
MENTAL ALERTNESS 27
MILITARY 7, 10, 49, 56-58, 60, 58
MILITARY DISCHARGE 49
MONTHLY TRAINING 20, 22
MOTIVATION 12, 59, 62, 127
MOTIVES 157
MULTIPLE CHOICE EXAMS 71
MULTIPLICATION PROBLEMS 191

N
N.F.P.A. 17, 366, 368, 371
N.F.P.A. STANDARD #10 371
NATIONAL FIRE PROTECTION ASSOCIATION
    N.F.P.A. 366

N
(continued)

NET PUMP PRESSURE	271, 359
NEUTRONS	304
NON-STRESSFUL ORAL	156
NOZZLE REACTION	284
NOZZLE PRESSURE	284

O

OBJECTIVES	57
OCCUPANCIES/CLASSIFICATION	368
OCTANE	311
OHMS	303
OIL	309, 310, 312
DILUTION	309
GAUGE	309
S.A.E	309
VISCOSITY	309
OPENINGS	368
OPPORTUNITIES	18
ORAL BOARDS	156, 162, 156
ORAL INTERVIEW	2, 11, 39, 40, 49, 61, 65, 153-158, 163, 164, 174, 155, 157, 158, 175
ORAL INTERVIEW PREPARATION	157
ORAL INTERVIEW QUESTIONS	164, 174
ORAL INTERVIEWS	156
ORDERS:	
INCONSISTENT	101, 111
ORIENTATION	36
OS&Y VALVE	376
OVERHAUL	21, 25, 69, 171, 288, 364, 25, 364

P

PACKING	361
PAID FIRE DEPARTMENT	3
PAID FIREFIGHTER	3, 60
PANIC HARDWARE	368
PARALLEL	361
PARAMEDIC TRAINING	58
PERCENTAGE PROBLEMS	195
PERFECT FIREFIGHTER	43
PERFORMANCE	10, 11, 16, 20, 21, 27-30, 157, 224, 250, 302, 363, 27
PERFORMANCE EVALUATION	29
PERSONAL APPEARANCE	12, 28, 29, 163
PERSONAL DATA	57, 59
PERSONAL HYGIENE	7
PERSONAL INFORMATION	55, 60
PERSONAL SAFETY	22
PERSONNEL DEPARTMENTS	41, 44, 41
PHYSICAL FITNESS	6
PHYSICS	69, 93, 225, 277-279, 285, 288, 289, 298, 299, 318, 285
PICKHEAD AXE	305
PIKE POLE	305
PISTON	308
PITOMETER:	306
POLICY	21
POSITION OF FIREFIGHTER	8, 13, 39, 55, 56, 63, 154, 155, 162, 164, 166, 173
POST INDICATOR VALVE:	
PIV	375
POWER EQUIPMENT	23
PRACTICAL EXAM	21
PRACTICE	157

P
(continued)

PRACTICE EXAMS	40, 157, 39
PRACTICE ORAL INTERVIEWS	174
PRE-READING	225-227, 238, 275
PREPARATION MANUALS	78
PRESSURE	361
BACK	284
KINETIC	287
VAPOR	281
ATMOSPHERIC	284
NOZZLE	284
PROBATION	8, 10, 20, 28, 20
PROBATION PERIOD	28
PROBATIONARY FIREFIGHTERS	20
PROBATIONARY REQUIREMENTS	20
PROCEDURE	21
PROGRESSION	77
PROPORTION	10, 12, 80, 88, 97, 176, 181, 183, 309, 310, 373
PROPORTIONER	373
PROSPECTIVE FIREFIGHTER	8, 18, 39, 18
PROTONS	304
PUBLIC RELATIONS	6, 11, 13, 18, 28, 29, 369
PULLEYS	90, 92, 97, 302, 314, 319, 326, 345
PUMP	
CENTRIFUGAL	359
CERTIFICATION TEST	362
DELIVERY TEST	363
IMPELLER:	363
MULTI-STAGE	363
POSITIVE DISPLACEMENT	358, 360
SERVICE TEST	363
PUMP CAVITATION	361
PUMP OPERATOR	4, 19
PUMPER	4, 19, 26, 64, 90, 180, 221, 251, 271, 272, 284, 310, 358, 359, 361-363, 271, 310, 359
PUMPER/QUAD	359
PUMPER/QUINT	359
PUMPER/TRIPLE	271, 359
PUMPING OPERATIONS	251, 252
PUMPS	361
ACTION	360
CAPACITIES	271
CENTRIFUGAL	269, 359, 360
DISPLACEMENT	358
POSITIVE	358-360
POSITIVE-CAPACITY	358
PRINCIPLE	359
RECIPROCATING	358
TESTING	359
BOOSTER	362

Q

QUAD	359
QUALIFICATIONS	8, 9, 12, 17, 18, 36, 48, 49, 55, 154, 155, 157, 161, 164, 9, 36
QUESTIONS:	
ADDITION	187
DIVISION	193
MECHANICAL OBJECTS	351
MULTIPLICATION	191
PERCENTAGE	195
READING	240

Q
(continued)

QUESTIONS (continued)
- SCIENCE 289
- SITUATION 99
- SPELLING 132
- SUBTRACTION 189
- PULLEYS, GEARS, LEVERS 319

QUINT 359

R

- RADIATOR 309
- RADIATOR COOLER 359
- RADIUS 285
- RATIOS 180, 181, 240, 241
- READING 76
- READING COMPREHENSION .. 68, 78, 80, 224, 226, 228, 233, 236, 237, 239, 275
- READING CONCENTRATION 236, 237
- READING MEMORY 238
- READING QUESTIONS 240
- RECORDS 17, 27
- RECTANGLES 209, 210
- RECTIFIER 308
- REFERENCES 56, 58, 60, 61, 58
- RELATED WORK 17
- RELAY PUMPING 253
- RELIEF VALVE 358
- REPORT OF PERFORMANCE 29
- REPORTS AND RECORDS 27-29, 27
- REQUIREMENTS 9, 36
- REQUIREMENTS AND QUALIFICATIONS 8, 9, 48, 49
- RESIDENCE 12
- RESPOND 18
- RESPONSE TO ORDERS 28
- RESUME 55, 56
- RESUME EXAMPLE 58-61
- RIGHT TRIANGLE 211, 221
- ROMAN NUMERAL 217
- ROOKIE FIREFIGHTER .. 41, 101, 104-106, 111, 115, 116, 118
- ROPES AND KNOTS 21, 24

S

- SAFETY PROCEDURES 27
- SALVAGE 25, 364
- SALVAGE COVERS 365
- SAMPLE TEST 76
- SCBA 306
- SCIENCE 36, 39, 42, 57, 58, 60, 61, 68, 215, 230, 277-279, 285, 288, 293, 306, 318
- SELF ASSESSMENT 36, 157, 36
- SERVICE TEST 362, 363
- SHEEPSHANK 350
- SITUATION QUESTIONS 79, 99, 170
- SIZE-UP 73, 259, 73, 259
- SMOKE DETECTORS 370
- SPECIAL REQUIREMENTS 11, 41, 11
- SPECIFIC GRAVITY 93, 240, 280, 285, 307, 308, 240
- SPECIFIC HEAT 93, 94, 98, 279
- SPECIFICATIONS 36
- SPEEDOMETER 309
- SPELLING 48, 68, 132, 137-140
- SPELLING TEST 132, 137-139
- SPONTANEOUS IGNITION 258

S
(continued)

SPRINKLER
- DELUGE 376
- DRY PIPE 376
- PRE-ACTION 376
- RATINGS 375
- WATER SUPPLY 375
- WET PIPE 376

- SPRINKLER COVERAGE 375
- SPRINKLER PATTERN 376

SPRINKLER SYSTEMS:
- OUTSIDE 376
- AUTOMATIC 376

- SQUARE MEASURE 216
- SQUARES 209, 210
- STANDARDS OF PERFORMANCE 27

STANDPIPE SYSTEMS:
- TYPES 376
- CLASS I 377
- CLASS II 377
- CLASS III 377

- STATIC PRESSURE 376
- STATION WORK 27, 29, 27, 29
- STEAM 372
- STRATEGY 25, 28, 58, 61, 225, 267
- STRATEGY AND TACTICS 25
- STRESSFUL ORAL 156
- STRESSFUL ORAL BOARDS 156
- STUDY HABITS 34, 35
- SUBJECT MATTER 68, 71, 78, 225
- SUBLIME 279
- SUBTRACTION PROBLEMS 189
- SUCCESS 36
- SUPERVISION 4, 36, 363
- SUPPLEMENT 54
- SUPPORT SERVICES 5
- SURFACE TENSION 289, 372

T

- TACHOMETER 309
- TACTICS 25, 58, 61, 225, 267, 25
- TARGET HAZARD 367, 369
- TEAM WORK 6, 27-29, 273
- TECHNICAL SKILLS ... 56, 57, 59, 57, 59

TEST
- APTITUDE 99
- VOCABULARY 120, 126
- PREPARATION 78
- SPELLING 138

- TESTING 21
- THE ORAL INTERVIEW   2, 11, 39, 40, 49, 65, 154, 155, 157, 158, 163, 174
- THE WRITTEN EXAMINATION   68, 79, 155
- THERMOSTAT 286, 309
- TILLERMAN 312

TIRES
- UNBALANCE 312
- WEAR 75, 312

- TOE-IN 312
- TOOLS 305
- TOOLS: MATCHING QUESTIONS 348
- TOOLS AND APPLIANCES 27
- TOOLS AND EQUIPMENT 13, 19, 27, 68, 69, 225, 244, 271-273, 302, 305
- TOOLS: FORCIBLE ENTRY 364
- TORQUE 310
- TRAINEE 3, 12

T
(continued)

TRAINING 3, 5, 6, 9-11, 13-16, 18, 20-22, 25, 26, 28, 36, 42, 49, 55, 57-61, 64, 65, 100, 111, 127, 141, 155, 157, 244, 262, 10, 21, 36
TRAINING DIVISION 5
TRAINING SCHEDULE 20
TRANSFER VALVE 361
TRANSFORMER 286, 290, 304, 313
TRANSMISSION 309
TRIANGLE 74, 93, 98, 211, 221, 274, 371
TRIPLE COMBINATION PUMPER 271, 272, 359
TROY WEIGHT 216
TRUE FALSE QUESTIONS 72
TWO STAGE PUMP 361
TYPES OF EXAMS 68, 71
TYPES OF ORAL 156
TYPICAL DUTIES 13

U

UNIFORM BUILDING CODE
  U.B.C. 366
UNIFORM FIRE CODE
  U.F.C. 366

V

VACUUM 256, 269, 284, 287, 312, 358-360, 362
VALVERANSFER 361
VALVE LIFTER 309
VAPOR DENSITY .. 93, 98, 241, 279, 241
VAPOR LOCK 311
VAPOR PRESSURE 281
VAPORIZATION 73, 241, 261, 281
VELOCITY 361
VENTILATION 13, 21, 25, 69, 103, 111, 114, 254, 258, 288, 293, 309, 364, 365, 369, 25, 111, 254, 364
VERBAL ABILITY 120
VISITING FIRE STATIONS 37, 38, 37
VOCABULARY TEST 120, 125, 126, 128-131
VOLTAGE . 286, 303, 304, 307-309, 313
VOLTAGE REGULATOR 308
VOLUME/CAPACITY 210
VOLUMES 207-209, 314, 360
VOLUNTEER FIRE DEPARTMENT 3
VOLUNTEER FIREFIGHTER 3, 57, 60
VOLUTE 270, 360, 361, 363, 360

W

WAGES 7
WAIVERS 12
WALLS 367
WATER 5, 13-16, 19, 25, 37, 72, 73, 76, 77, 90, 91, 93, 94, 97-99, 102, 103, 106, 107, 109, 112, 113, 117, 118, 119, 148, 151, 213-215, 218, 221, 240, 245, 247-254, 256, 259-261, 263-265, 268, 269, 270-272, 279-284, 286, 287, 289-294, 296, 303-305, 307-310, 343, 346, 347, 358-363, 365, 371-373, 375, 376, 283, 371, 373
WATER: LIGHT 373
WATER:VISCOUS 373
WATER CONSUMPTION .. 215, 307, 308
WATER ENGINEERING 5
WATER EXTINGUISHER 263
WATER FOG 372
WATER HAMMER 250, 252
WATER SUPPLY 5, 15, 19, 25, 37, 73, 77, 221, 250, 259, 376, 25
WET-WATER 372
WETTING AGENTS 372
WHEELS . 283, 292, 302, 309, 312, 314, 330-332, 334, 335, 340
WINDWARD: 365
WOMEN 11
WORD PROBLEMS 176, 197, 205
WORK ENVIRONMENT 68, 165
WORK EXPERIENCE 59, 60, 63, 59
WORK HABITS 35, 65
WORKING CONDITIONS 6
WORKING HOURS 6, 27, 29
WRITTEN DIRECTIONS 48
WRITTEN EXAM 70
WRITTEN EXAMINATION 68, 79, 155

FIRE SERVICE BOOKS:

BOOK #1

"FIRE ENGINEER WRITTEN EXAM STUDY GUIDE": 180 page book with ten chapters containing over **2500** selections of information that **ALL FIRE FIGHTERS** should know. Includes the following subject matter: Fire apparatus, tools and equipment, fire streams, fire pumps, fire hydraulics, fire extinguishing systems, fire behavior water supply, hazardous materials, and fire prevention.(2nd ed.) **PRICE: $15.95**
ISBN 0-938329-52-9 LCCN 86-81239

BOOK #2

"FIRE ENGINEER PRACTICAL - ORAL EXAM STUDY GUIDE": 192 page book with seven chapters containing over **400 PRACTICAL - ORAL INTERVIEW** and **SITUATION** type questions. Each question is followed by a **SUGGESTED** response. Includes the following subject matter: Oral interview preparation, job knowledge, general knowledge, personal information, along with actual situation questions. (2nd ed.) **PRICE: $15.95**
ISBN 0-938329-53-7 LCCN 88-80888

BOOK #3

"FIRE CAPTAIN WRITTEN EXAM STUDY GUIDE": 288 page book with ten chapters containing over **3000** selections of information that **ALL FIRE FIGHTERS** should know. Includes the following subject matter: Fire administration, leadership, training, fire prevention, fire behavior, fire apparatus and equipment, fire extinguishing systems, fire streams, water supply, and hazardous materials. **PRICE: $18.95**
ISBN 0-938329-54-5 LCCN 88-80890

BOOK #4

"FIRE CAPTAIN ORAL EXAM STUDY GUIDE": 228 page book with ten chapters containing over **400 ORAL INTERVIEW** and **SITUATION** type questions. Each question is followed by a **SUGGESTED** response. Subject matter includes: Oral interview preparation, job/general knowledge, personal information, and actual situation questions. This book includes Assessment Center and Simulator exam information. **PRICE: $18.95**
ISBN 0-938329-55-3 LCCN 88-80889

BOOK #5

"THE COMPLETE FIREFIGHTER CANDIDATE": 150 page book with eight chapters containing the **MOST COMPLETE** inventory of information available. This book will guide Firefighter candidates through all the key steps that need to be taken in order to secure the position of **FIREFIGHTER**. Includes the following subject matter: Introduction to the Fire Service, Fire Department familiarization, exam check list, exam examples, how to locate exams, job announcements, job applications, resumes, the exam process, exam divisions, what to prepare for, such as: Oral-written-agility-and medical exam information. **PRICE: $12.95**
ISBN 0-938329-58-8 LCCN 89-81738

BOOK #6

"FIREFIGHTER WRITTEN EXAM STUDY GUIDE": 336 page book with twelve chapters containing over **3000** selections of information - questions - answers that **ALL PROSPECTIVE FIREFIGHTERS** should know. This book is the most complete book of its type. **LOADED WITH INFORMATION!** This book covers all types and catagories of Entrance Firefighter **WRITTEN EXAMS**. **PRICE: $19.95**
ISBN 0-938329-59-6 LCCN 89-81736

BOOK #7

"FIREFIGHTER ORAL EXAM STUDY GUIDE": 226 page book with eight chapters containing over **400 ORAL INTERVIEW** and **SITUATION** type questions. Question are followed by a **SUGGESTED** response. This book covers all phases and catagories of the Entrance Firefighter **ORAL EXAM**. **PRICE: $15.95**
ISBN 0-938328-61-8 LCCN 89-81737

The perfect bound soft cover editions are available from : **INFORMATION GUIDES**, P. O. BOX 531, HERMOSA BEACH, CA 90254. **PHONE : (213) 379 - 1094** or **TOLL FREE : 1 - 800 - "FIRE BKS"**

FIRE SERVICE BOOKS:
(CONTINUED)

BOOK #8

"A SYSTEM FOR ADVANCEMENT IN THE FIRE SERVICE": 170 page book with six chapters, containing all of the essential procedures and steps that Firefighters should follow in order to **PROMOTE** within the **FIRE SERVICE**. This is the only book available with this type of information! **PRICE: $9.95**
ISBN 0-938329-56-1   LCCN 88-83385

BOOK #9

"EMERGENCIES IN THE HOME": 150 page book with 8 chapters containing the basic information that **YOU** and your **FAMILY** should be aware of during emergency situations. The information included in this book will assist **YOU** and your **FAMILY** in surviving various types of emergencies, such as 1. First Aid: breathing, drowning, broken bones, bleeding, seizures, burns, electrical shock, poisoning, etc. 2. Fires: escape plans, preparedness before, during and after. 3. Earthquakes; preparedness before, during and after. 4. Floods: preparedness before, during and after. 5. Weather Emergencies: preparedness before, during and after. 6. Police emergencies: preparedness before, during and after. 7. Emotional Crisis: preparedness before during and after.   **PRICE: $16.95**
ISBN: 0-938329-74-X   LCCN 91-073746

BOOK #10

"FIRE SERVICE ENTRANCE EXAM PREPARATION" - Instructor Manual: 300 page book with **15 LESSON PLANS** that are to be used by the Fire Technology Instructor in conjunction with the class: **FIRE SERVICE ENTRANCE EXAM PREPARATION**. The lesson plans cover all of the essential material required to guide and instruct Firefighter candidates to a career in the Fire Service. The information includes Fire Service Orientation, the Application Process, the Written and Oral Exam. The written exam information includes: math, reading comprehension, chemistry, physics, and mechanical comprehension. The oral exam information includes questions with suggested responses along with practice sessions and evaluations.
PRICE: $35.00
ISBN 0-938329-72-3   LCCN 91-092875

BOOK #11

"FIRE SERVICE ENTRANCE EXAM PREPARATION" - Student Manual: 400 page book with 11 chapters this book is to be used by Fire Technology Students in conjunction with the class: **FIRE SERVICE ENTRANCE EXAM PREPARATION**. The chapters cover all of the essential material required to guide Firefighter candidates to a career in the Fire Service. The information includes Fire Service Orientation, the Application Process, the Written and Oral Exam. Written exam material includes: math, reading comprehension, chemistry, physics, and mechanical comprehension. Oral exam material includes questions with suggested responses.   **PRICE: $39.00**
ISBN 0-938329-71-5   LCCN 91-073745

Cleverly drafted, each bit of information is presented in a new and unique method for efficient and organized studying. The locating of all the various reference material is often the most difficult and time consuming portion of the study process. The eleven books described above are designed to reduce the time of this process along with increasing the quality of **YOUR** study sessions. BOOK #9 is not an exam study guide but it is an easy reference of emergency information that should be studied prior to emergencies that may take place in our daily lives.

Every Firefighter and/or Firefighter candidate that has been or is currently involved in the Fire Service exam process will attest to the need for books of this type. Firefighters and/or Firefighter candidates that are or will be preparing for Fire Service exams will have a specific need for these books!

The perfect bound soft cover editions are available from : **INFORMATION GUIDES**, P. O. BOX 531, HERMOSA BEACH, CA 90254. **PHONE** : (213) 379 - 1094 or **TOLL FREE** : 1 - 800 - "FIRE BKS"

BOOKS AVAILABLE FROM INFORMATION GUIDES:
1-800-"FIRE BKS"

BOOK # 1: "FIRE ENGINEER WRITTEN EXAM STUDY GUIDE", 2nd ed.	$15.95
BOOK # 2: "FIRE ENGINEER ORAL EXAM STUDY GUIDE", 2nd ed.	$15.95
BOOK # 3: "FIRE CAPTAIN WRITTEN EXAM STUDY GUIDE"	$18.95
BOOK # 4: "FIRE CAPTAIN ORAL EXAM STUDY GUIDE"	$18.95
BOOK # 5: "THE COMPLETE FIREFIGHTER CANDIDATE"	$12.95
BOOK # 6: "FIREFIGHTER WRITTEN EXAM STUDY GUIDE"	$19.95
BOOK # 7: "FIREFIGHTER ORAL EXAM STUDY GUIDE"	$15.95
BOOK # 8: "A SYSTEM FOR ADVANCEMENT IN THE FIRE SERVICE"	$ 9.95
BOOK # 9: "EMERGENCIES IN THE HOME"	$16.95
BOOK #10: "FIRE SERVICE ENTRANCE EXAM PREPARATION", INSTRUCTOR MANUAL	$35.00
BOOK #11: "FIRE SERVICE ENTRANCE EXAM PREPARATION", STUDENT MANUAL	$39.00

BOOKS ARE ALSO AVAILABLE FROM:

PERFECT FIREFIGHTER---------1-800-326-8401
FIREFIGHTERS BOOKSTORE-1-714-536-4926
KEAU HOU BOOKS---------1-808-322-8111

EDCON PRESS---------1-800-253-3266
GALL'S FIRE BOOKS-1-800-524-4255
LOU'S BOOKS---------1-213-428-5002

BOOK REVIEWS:

FIRE CHIEF MAGAZINE: "Information that Fire Fighters should know in preparing for Fire Service exams..."

FIREHOUSE MAGAZINE: "Well thought-out study books for the person aspiring to a career in the Fire Service..."

REKINDLE MAGAZINE: "Information is presented in a new and unique method for efficient and organized studying..."

FIREFIGHTER "NEWS": "Every Firefighter or Prospective Firefighter currently involved in the Fire Service exam process will attest to the need for books of this type..."

ORDER FORM:

"RUSH" ORDERS PHONE:
(213) 379-1094 OR CALL TOLL FREE: 1-800-FIRE BKS = 1-800-347-3257

INFORMATION GUIDES, P.O. BOX 531, HERMOSA BEACH, CA 90254

BOOK #1_____ COPIES @ $15.95 PER COPY $_____
BOOK #2_____ COPIES @ $15.95 PER COPY $_____
BOOK #3_____ COPIES @ $18.95 PER COPY $_____
BOOK #4_____ COPIES @ $18.95 PER COPY $_____
BOOK #5_____ COPIES @ $12.95 PER COPY $_____
BOOK #6_____ COPIES @ $19.95 PER COPY $_____
BOOK #7_____ COPIES @ $15.95 PER COPY $_____
BOOK #8_____ COPIES @ $ 9.95 PER COPY $_____
BOOK #9_____ COPIES @ $16.95 PER COPY $_____
BOOK #10_____ COPIES @ $35.00 PER COPY $_____
BOOK #11_____ COPIES @ $39.00 PER COPY $_____

SHIPPING AND HANDLING:
BOOK RATE = $2.00 1ST COPY ADD : 50 CENTS PER EACH ADDITIONAL COPY ORDERED. "RUSH" FIRST CLASS MAIL = $3.50 1ST COPY ADD $1.00 FOR EACH ADDITIONAL COPY.

SHIPPING AND HANDLING = $_____

CALIFORNIA SALES TAX:
CALIFORNIA RESIDENCE, ADD 7% SALES TAX = .07 TIMES TOTAL AMOUNT FOR BOOKS.

MAKE CHECK OR MONEY ORDER OUT TO: "INFORMATION GUIDES"

TAX = $_____

TOTAL AMOUNT ENCLOSED FOR ORDER = $_____

NAME_____
FIRE DEPT._____
ADDRESS_____
CITY_____
STATE & ZIP_____

CONVENIENT CREDIT CARD ORDERS:
VISA_____ MASTER CARD_____
CARD#_____
EXPIRATION DATE_____
NAME ON CARD_____
SIGNATURE_____

IF NOT FULLY SATISFIED, BOOKS MAY BE RETURNED FOR A FULL REFUND!